A Changing Climate for Science

Sophie C. Lewis

A Changing Climate for Science

palgrave
macmillan

Sophie C. Lewis
Fenner School of Environment and Society
The Australian National University
Canberra, Australia

ISBN 978-3-319-54264-5   ISBN 978-3-319-54265-2 (eBook)
DOI 10.1007/978-3-319-54265-2

Library of Congress Control Number: 2017936879

Cover image: Modern building window © saulgranda/Getty

Printed on acid-free paper

This Palgrave Macmillan imprint is published by Springer Nature
The registered company is Springer International Publishing AG
The registered company address is: Gewerbestrasse 11, 6330 Cham, Switzerland

Preface: An Apprenticeship

I've wanted to be a scientist since I was a small child. This book tells my story of becoming a scientist, and of struggling to reconcile this journey with my experience as a climate scientist. This is also my story of carving out a new view of science, and eventually of coming to identify myself as a postmodern scientist. To some, this term seems senseless at best, oxymoronic at worst; my aim throughout is to make the seemingly senseless become useful.

My story began in 1986 when my parents took me stargazing as a young child in the hope of glimpsing Halley's Comet. We trudged for some time through the open grassy fields and then we waited, and we waited. It was a pale, grey night. In our part of the world, thick banks of stratus cloud masked the comet's infrequent voyage across the skies. There was nothing to be seen that night, but still I was thrilled. I didn't know it at the time, simply and childishly excited, but my interest had been piqued by science.

My family spent a lot of time in the foothills of the Australian Alps. As a child in the vast eucalyptus bush I collected furiously—old bones or teeth, snake skins, tadpoles, feathers, leaves, seed pods, river stones, freshwater yabbies, anything mobile and anything sessile. My uncle gave me a microscope and slide-making kit, and then a few years later, my grandmother gifted me a small telescope. In my mind, a rock might reveal a fossil and a starry night might give up a particularly breath-taking meteorite. I was hungry for answers to questions, hungry for new knowledge. The world of my childhood was a place to be consumed one piece at a time, all in quick succession.

At high school I studied as much science and mathematics as I could, and later when I started university, the teaching and learning of science became increasingly formalised. A small pocket full of seedpods and beetle casings was replaced by notebook sketches of Bunsen burners, atomic models, calculus, taxonomic ranks, fluid mechanics, general relativity and number theory. I still collected furiously, but now instead of odd bits and pieces found in gullies and frost hollows, I gathered information, consuming and ordering facts of ever-greater complexity. At my university, as the years of an undergraduate degree are completed, standard coursework characterised by dense discipline-specific information, models and methods, slowly gives way to specialised research training. Philosophers of science Thomas Kuhn and Karl Popper are exalted, the scientific method memorised, logic discussed, and inductive and deductive reasoning delineated.

After I finished my undergraduate degree, I eagerly signed up to an Honours year. I decided to specialise in palaeoclimatology, which is the study of past climate change. Palaeoclimatology was located in a notoriously ambiguous Geography department, where I found myself one of only two young physical geographers-in-training amongst a large group of human geography students. By some peculiar university bureaucratic hurdle, all students were required to participate in the same units of Honours coursework, which were impossibly tasked to prepare us equally well as researchers in disparate fields. We all learned about the twentieth-century turns in the social sciences that influenced thinking in the discipline of geography, as well as the requirements of a scientist.

During the early days of that geography course, our professor asked who believed in the idea of an absolute truth. Put on the spot and filled with the signature undergraduate fright of being asked to think or act, no one raised a hand. The professor went on to chide the small band of physical geographers as poor specimens of scientists. What kind of scientist doesn't uphold the idea of a universal, discoverable truth? I'm quite sure that I went on to hastily note down that a scientist believes in a singular understanding of the world. Quick! Sophie! That's a scientist! Be that, do that!

Learning alongside human geography students, I superficially digested positivism, Marxism, structuralism, and postmodernism without any meaningful understanding of what each entailed. 'No matter,' I thought! These social scientific curiosities were merely curious. The course was probably a useful experience, and I would at least know some more words that I could repeat in conversation later to seem nonchalantly well read and eloquent. And when the semester snapped shut with our final exams, I would at long last be a scientist! As a 'proper' scientist, my work was undeniably more important than the rest of my cohort, with their vague, poorly defined theories and methods that characterise research in social science.

After I submitted my Honours thesis, I moved to a new university to earn my PhD by researching a slightly different type of palaeoclimatology. A traditional understanding of the PhD is as an apprenticeship in science. A scientist is differentiated from a non-scientist—a non-expert—by these years of apprenticeship to an erudite supervisor, which essentially constitutes a specialised training. A PhD is the standard process through which students are metamorphosed into scientists. When I finished my PhD, I soon began a research fellowship in climatology, which was followed quickly by a second and third fellowship. I had trained to become a scientist and was finally there. Little Sophie would be so pleased!

Throughout my informal and formal education, I loved science and yearned to be a scientist. Yet it had never occurred to me to ask—what is science and what is a scientist? If I recall my rote-learned course material from those hazy, fun-filled undergraduate days, science constitutes a system of knowledge. It is a systematic enterprise that obtains knowledge through a formalised approach called 'the scientific method'. A hypothesis is posed and tested, and knowledge acquired, but not produced. Science must be reproducible and it must be falsifiable. This was the singular epistemology that defined a scientist, as simply one who enacted science using these methods. That Honours' level lecture on the central idea of an absolute truth remains the last formal discussion I've had about science and its ways of knowing. These helpful guardrails remain in place to stop scientists veering beyond this understanding.

After I commenced my first real scientific job as a postdoctoral research fellow, I began to feel vaguely uneasy about my research as a scientist. This uneasiness stubbornly refused to pass. It turns out that in my field of research, contemporary approaches often do not subscribe to the techniques or methodologies described by my undergraduate training. For example, I spent four years of my PhD reconstructing past changes in climate from incomplete data sources that lend themselves wonderfully to plural interpretations. My research now aims to understand current changes in climate using complex computer climate models that we are possibly unable to falsify.

Is this still science? If my scientific data are not readily reproducible, do they remain useful? Or what if we imagine that rather than positing and testing hypotheses, I generate novel understandings of the world by haphazard data mining? Is this then inherently 'unscientific'? Where does this leave scientists? And, crucially, where does this leave science? With great dedication, I dutifully did my apprenticeship, but did I become an actual scientist?

In this book, I propose a new view of science. This is my own reappraisal of science, a re-imagining of scientific practise as nuanced, transparent, diverse and creative. Ultimately, I pose a place *beyond* current understandings of science's ways of knowing. I describe this as a 'hinterland,' a conceptual space that allows for diverse practices of science, centred on a flexible and inclusive way of being a scientist. Within this hinterland, I describe myself as a new type of scientist by using the seemingly oxymoronic description of 'postmodern scientist.'

As a caveat, I do not profess to have a deep understanding of theory of knowledge. A philosopher of science or a sociologist could address these questions with far more intellectual heft than I can. As such, the following chapters are not deeply rooted in literature. Instead, I describe my own experience of grappling with myself over whether I am a scientist, and eventually coming to reject the universal utility of the narrow approaches that I rote learned as an 'apprentice' scientist. These insights are simply one scientist's thoughts about being a particular kind of scientist.

Finally, a word about what this book is *not*. This book is not a negative appraisal of science, climate science or climate scientists. It is my *experience* of science, climate science and climate scientists. In many cases in the following chapters, I highlight particular examples in the literature, or commentaries, but I emphasise that this is not because I view these studies as wrong, or poor, or ill-considered. It is quite the opposite; I present these as examples of valuable contributions to our understandings of the discipline and explore these specifically to demonstrate that science does not exhaust all knowledge. I discuss this literature in good faith, as a member of this community and as a committed climate scientist. In doing so, I hope that these explorations will be viewed as such, as an affirmative critique.

Acknowledgements

The ideas of this book were nurtured by many conversations with friends and colleagues, which were, in equal measures, exciting, irritating, uncomfortable and energising. For this, I am humbly grateful to my colleagues. I am grateful for the lively and generous intelligence of my students and my young colleagues at the University of Melbourne, The Australian National University and my collaborators at the University of New South Wales and Monash University for indulging long discussions on the fringes of our work.

I am equally grateful for the kindness of my senior colleagues in indulging my interests in these fringes of our discipline. While many young researchers are bound so tightly by the realities of precarious funding and uncertain salaries that they achieve scientific greatness at the expense of freedom of thought, I have been given every opportunity and encouragement to pursue my own interests through my ARC DECRA funding (DE160100092). For this opportunity, I offer my gratitude to the visionaries who lead the Australian Research Council funded Centre of Excellence for Climate System Science.

I would particularly like to thank the provocative Julia Jasonsmith, for providing a decade of alternative and challenging viewpoints, and Marie-Louise Ayres, for her eagle-eyed assistance. Finally, I am indebted to my best friend, Cathy Ayres, whose presence fills me with the gumption to try. She provided endless enriching comments and endless enriching encouragement on the regular occasions my courage to pursue new ideas faltered.

CONTENTS

LIST OF FIGURES

CHAPTER 1

The Want of Any Name

Abstract Lewis explores science as a way of knowing that occurs through implementing the scientific method. Through the example of pioneering physician Edward Jenner and his development of the smallpox vaccination, Lewis outlines how the scientific method is applied. She explores the key processes of science, including hypotheses, observations and theories, and the key concepts of science, including falsifiability, repeatability and objectivity. Lewis contrasts this orthodox understanding of science with a critical reflection on her own experience of training to be, and practicing as, a climate scientist. She exposes a gulf between the way that science is traditionally perceived and commonly described, and the way it is actually conducted in contemporary disciplines. Lewis provides a much-needed exploration of modern scientific practice through a focus on climate science.

Keywords Scientist · Scientific method · Theory · Scientific training · Assemblage of concepts

A surprising account of an important series of discussions was published in 1834. Deliberations were prompted by the realisation of the 'want of any name' by which 'students of the knowledge of the material world' could collectively be described (Anon 1834).

Philosophers was felt to be too wide and lofty a term, and was properly forbidden them by Mr Coleridge, both in his capacity as philologer and metaphysician; savans was rather assuming, besides being French instead of English.

It turns out 'some ingenious gentleman proposed that, by analogy with *artist*, they might form *scientist*.' Back in 1834, the proposed term of scientist was 'not generally palatable.'

Now the name scientist is so fitting that it seems strange and almost quaint to consider that this word was ever actively *decided* upon. The deliberate selection of the word 'scientist' to describe those who pursued science is interesting, and not just for the delightful idea of this tense, drawn-out search for the truly appropriate word to describe such intellectual activities. Prior to 1834, in the English language at least, there was little to group scientists together and hence little to distinguish scientists from non-scientists. Adopting a name signified an important change. The use of a name implies something collective, a shared endeavour, a systematic enterprise, mutual goals and connected values.

What is science, this endeavour that scientists have shared since 1834? The simplest answer, perhaps, is that science is a system of knowing. It is an enterprise characterised by its methodology, the **scientific method**. The scientific method in turn is an empirical approach that distinguishes the knowledge claims of science from the dubious, unfounded claims of **pseudo-science**. The scientific status of a theory is determined by its **falsifiability**, refutability or its testability. In *Conjecture and Refutations,* philosopher of science Karl Popper (1963) describes the search for truth as our greatest motivation in scientific discovery. The knowledge claims of science can be described as 'truthful' only through applying the principle of falsificationism. By this principle, *any* contradictory instance to a scientific theory is sufficient to falsify that theory, regardless of how many positive examples appear to support it. Hence the scientific method is the process that codifies scientific knowledge. This describes an **inductive** process of observation, hypothesis, testing and theory making, such that when scientists obtain consistency in **observation** and prediction, a **hypothesis** becomes a **theory**. Unlike the vernacular use of the word 'theory' to describe simply a hunch, or an idea, a scientific theory describes a consistent, coherent framework through which scientists can understand observed phenomena. A theory provides the framework for explaining observations and for making testable predictions.

How does this method work in practice? Let's take a look at the example of the work of Edward Jenner (1798). A physician and scientist,

Jenner is credited with having saved more lives than any other individual in history. He pioneered the world's first vaccine, successfully inoculating against smallpox by following this method:

Step 1. Edward Jenner astutely *observed* that local milkmaids were generally immune to deadly smallpox.

Step 2. He *hypothesised* that pus in the blisters that milkmaids received from contracting cowpox somehow protected them from the far more virulent smallpox virus.

Step 3. In 1796, Jenner tested his hypothesis. To do so, he scraped pus from a milkmaid's cowpox blisters, made a few scratches on a local boy's arms and rubbed some of the pus into the open scratches. The boys became mildly ill with cowpox but recovered rapidly. Most importantly, the child was not afflicted when exposed weeks later to smallpox.

Step 4. Jenner followed up with many experiments and some years later, in 1798, he published his research in his book *An Inquiry into the Causes and Effects of the Variolae Vaccinae; a Disease Discovered in some of the Western Counties of England, Particularly Gloucestershire, and Known by the Name of The Cow Pox* (Jenner 1798). He had used the scientific method to develop a theory that cowpox protects against smallpox infection.

Science is an iterative approach—one of repeated observation, hypothesis, experiment, observation, hypothesis, experiment, observation and eventually theory. It is intuitive, to the point that we might extend our category of accomplished scientists to include babies, who regularly implement the scientific method to understand their world:

Step 1. Baby will question what happens when different food is dropped from a high chair.

Step 2. Baby will design an experiment and drop said food, while carefully observing the results.

Step 3. Baby observes that certain foods behave differently when dropped from a height (and, of course, adults will respond accordingly by picking up food).

Step 4. Baby forms a theory to provide a coherent framework about the response of food to being dropped from a height.

Simplistically, the scientific method is a formalised understanding of our natural approach to investigating our external world. To scientists, however, the scientific method represents far more; it underpins science. It acts to differentiate science from non-science, and to differentiate scientists from non-scientists. Under this paradigm, science is an objective pursuit, scientific knowledge is not intuitive and the scientific method is not prejudiced. That means that a scientific experiment is repeatable, regardless of the observer. By following Jenner's method, *anyone* should be able to test the efficacy of cowpox for preventing smallpox infections. Although Jenner was clearly an ingenious and industrious scientist, his theory has been accepted not because of his natural scientific skills or his authority, but rather it has been embraced because of the **repeatability** of his results.

During my university training, I diligently learned that the scientific method is the basis of science and permits us to be scientists. While this understanding of scientific knowledge pursued through a clean and clear methodology is compelling, it is an idealised view. Contemporary science is composed of many disciplines, with various approaches and goals. Some science remains highly empirical, while some science is theoretical. As a result, for many disciplines of science, there is an unavoidable gulf between the way that science is traditionally perceived and commonly described, and the way it is actually pursued. When I consider this traditional understanding of science, I cannot avoid confronting the idea that I have unwittingly become a scientist who isn't a **Scientist**. I am living this contradiction between how science is perceived and how it is pursued.

What follows are eight chapters that explore the implications of viewing science through this singular lens—fixed in place by the scientific method and braced by immutable pillars, such as by **knowability** and **objectivity**—and present a new view of science. I begin discussing the practice of science using an **assemblage of concepts** that are useful for understanding science. These key concepts—knowability, objectivity, legitimacy, credibility and authority—prompt a series of questions about the production of science. Each chapter addresses aspects of this assemblage by drawing on my own experiences as a practitioner of science, and by using the literature from the sociology of scientific knowledge to explore these experiences.

I begin in *Chapter 2* by exploring the scientific ideal of knowability, and the ways in which the uncertain, the unknowable and the ungraspable are central to scientific knowledge. In *Chapter 3*, I explore the concept of the legitimacy of scientific knowledge by discussing the concept of falsifiability and the demarcation of pseudo-science. Next, the key concept of credibility is explored in *Chapter 4* through the lens of my own PhD experiences. In *Chapter 5* I interrogate the ideas of authority and expertise in scientific knowledge claims. The discussion then opens out as I explore the role of science in society, critically examining recent controversies in climate science and exploring the idea of objectivity in science in *Chapter 6*. Here I introduce the idea of an epistemological hinterland—a space beyond the objective, where scientists openly acknowledge the culture, beliefs and plural practise of science. This hinterland is expanded in *Chapter 7* through a discussion of curiosity as a key element of scientific inquiry.

Finally, in *Chapter 8*, I turn my critical attention to examining my own scientific practices and propose an alternative set of key descriptors of scientific practice that fill this conceptual hinterland. I discuss a new narrative for science and a new understanding of being a scientist, presenting the seemingly uncomfortable concept of the postmodern scientist as a re-imagining of both science and scientists. In this final chapter, described not as a conclusion, but rather as an invitation, I offer a perspective of the discipline of climate science beyond that explored in the individual chapters. I liken the grand challenge that climate change presents to society with the grand challenge climate science presents to science. I argue that just as climate change give us a chance to re-imagine the future, the same can be said for climate science providing us an opportunity to re-imagine science.

Throughout, key terms are indicated by bold font and are discussed further in each chapter's glossary. Where necessary for clarity, I use the term Scientist to describe this orthodox understanding of science pursued through a singular methodology, and use this capitalisation to distinguish this narrow understanding from the broader term scientist, which encompasses a plurality of experiences and approaches. I will use this distinction to discuss my experience of conducting a particular type of science, climate science, which resides in this gulf. On a practical note, the material covered in the following chapters is often academic, intermingled with first-hand reflections of my experience of being a climate scientist. More scientifically trained readers might enjoy the description of the attribution of extreme climate events to human influences in *Chapter 2*, or the discussion of climate models and the utility of the field of palaeoclimatology in *Chapter 6*. Alternatively, general readers might prefer to skim over these and head to my reflections of my own research experiences detailed in *Chapters 4–6*.

GLOSSARY

Assemblage of concepts An array of related key scientific ideals that provide a useful basis for exploring, conceptualising and interrogating science as a mean of enquiry.

Falsifiability The knowledge claims of science are assessed through examining their falsifiability. *Any* contradictory instance to a scientific theory is sufficient to falsify that theory, regardless of how many positive examples appear to support it.

Hypothesis A hypothesis is a suggested solution for a phenomenon that is currently unexplained by scientific theory. There is no predetermined outcome of a hypothesis and it must be able to be supported or refuted through experimentation and observation.

Inductive Inductive reasoning takes specific information and makes a broader generalisation that is considered probable through strong evidence, though not necessarily proven.

Knowability The capability of being known, apprehended and understood.

Objectivity The ideal of objectivity in science is the idea that scientific claims and results are true outside of, and not influenced by, a scientist's individual biases, interpretations and perspectives.

Observation The act of receiving external knowledge of the world through one's senses, or by recording information through scientific instruments.

Pseudo-science Here I use a precise meaning of pseudo-science, beyond the popular categorisation of beliefs such as astrology and homeopathy, which are mistaken for science, but are actually pseudo-science. Here, pseudo-science is more broadly a collection of practices that do not follow the scientific method or cannot be falsified, and hence lack true scientific status.

Repeatability The close agreement between successive measurements or experiments carried out under identical conditions.

Scientific method An approach of systematic and repeated observation, measurement, experiment and the formulation, testing and modification of scientific hypotheses.

Scientist I use the term Scientist to describe this orthodox understanding of science pursued through a singular methodology, and use this capitalisation to distinguish this understanding from the broader term scientist.

Theory   A coherent group of propositions that explain a natural phenomenon and are confirmed through repeated experiment and observations.

REFERENCES

Anon (1834) On the connexion of the physical sciences. The Quarterly Review 51:54–59.

Jenner E (1798) An Inquiry Into the Causes and Effects of the Variolae Vaccinae: A Disease Discovered in Some of the Western Counties of England, Particularly Gloucestershire, and Known by the Name of the Cow Pox. printed for the author, by D.N. Shury, 1801. London: Daniel Nathan Shury.

Popper K (1963) Conjectures and Refutations: The Growth of Scientific Knowledge. London: Routledge & K. Paul.

CHAPTER 2

Neither Necessary Nor Sufficient

Abstract Lewis examines the key scientific concept of knowability using case studies from climate science. By unpacking the ideas of a settled climate science and scientific consensus, Lewis presents the uncertain and unknown as fundamental aspects of the world. Lewis discusses two disparate understandings of causality by exploring the attribution of extreme climate events to particular causes. The first is a mathematical framework of cause and effect, while the second embraces the capacities of both the human and the nonhuman to cause important effects as actants. Lewis explores these differing conceptualisations of causality through a critical reflection on her own scientific research, her proximity to a tragedy and her personal experiences of an extreme climate event. In response to these experiences, Lewis concludes with a recommendation around scientific terminology of causality.

Keywords Knowability · Causality · Attribution · Climate change · Extreme events · Nonhuman agency

The first entry in the assemblage of key concepts for science is **knowability**, which broadly refers to the idea that the physical world can be known, apprehended and understood. At first, any discussion of the known and unknown can easily become snarled in semantics or reduced to trivialities, as demonstrated by Fitch's Paradox of Knowability (2014). This logical

© The Author(s) 2017
S.C. Lewis, *A Changing Climate for Science*,
DOI 10.1007/978-3-319-54265-2_2

proof begins by succinctly stating that if we assume that every truth is knowable in principle, then every truth is actually known. As we evidently do not know all truths, there consequently exist unknowable truths. I do not plan to unpack the epistemological nuances of knowability, as such semantics only act to entangle discussions in logic, rather than to encourage reflection on scientific practice. Instead, I explore the implications of the limits of the known for the practice of science and the communication of scientific knowledge.

This chapter presents the unknowable, the uncertain and the ungraspable as fundamental aspects of the world. Despite this, the inevitability and importance of the unknown has not permeated the way that science is practiced and communicated. In climate science, findings such as the role of human activities in global warming are often described as 'settled.' In addition, the human and natural causes of destructive extreme weather events, such as floods and heatwaves, are described with certainty. Both examples conceal the ubiquity of the unknowable. In this chapter I demonstrate that subtle changes in language, framing and understandings allow the unknowable to fold into climate science without diminishing the value of scientific knowledge.

The Science Is Never Settled

The unknown is a fundamental aspect of the world. In his book *Ignorance* (2012), Professor Stuart Firestein, the Chair of the Department of Biological Sciences at Columbia University, details several well-known instances where knowledge has been shown to have limits. In these cases, the unknown is unavoidable. Firestein discusses physicist Werner Heisenberg's **Uncertainty Principle**, which tells us that at the subatomic level, we can never simultaneously know the position *and* momentum of a subatomic particle (as well as other such pairs of observations taken together). Such inescapable limits to our knowledge have been formally understood since at least the 1920s and 1930s when Heisenberg described uncertainty through his Principle. A key aspect of such limits to knowledge is that they are hard; meaning, for example, that a lack of instrumentation is not the reason we lack simultaneous measurements of subatomic positions and momentums. As Firestein (2012) summarises

> Heisenberg's result is not simply a case of lacking a good-enough measuring device. The very nature of the universe, what is called the wave-particle

duality of subatomic entities, makes these measurements impossible, and their impossibility proves the validity of this deep view of the universe. Some fundamental things can never be known with certainty. (p. 35)

Despite inherent uncertainty in understanding the world, a common refrain in discussions of human-caused climate change is the idea that the 'science is settled.' This phrase is shorthand for conveying that the vast majority of practicing climate scientists agree that **anthropogenic** greenhouse gas emissions, such as carbon dioxide, are largely responsible for the recently observed increase in global temperatures (IPCC 2013). The idea of settled science has primarily coalesced around a project focused on improving the communication of climate change science to the public, which quantified that 97% of practicing climate scientists agree on the fundamental science underpinning observed climate change (Cook et al. 2013). The '97% consensus' was subsequently used to demonstrate that climate change is not the contentious issue in the scientific community that it is widely believed to be.

However, this 'settled' idea has been appropriated both by advocates for aggressive cuts to greenhouse gas emissions, and also by sceptics who cite robust discussion in peer reviewed scientific literature to cast doubt on findings. Use of 'settled' terminology is widespread and expansive and has been extrapolated from first meaning that most climate scientists agree on the fundamental causes and aspects of climate change to meaning anything from 'the science behind *this* extreme weather event is settled' to 'the science of future climate projections is settled' to 'the science of the hiatus in global surface temperature is settled.' While 'settled science' is an effective approach for communicating the state of scientific understandings of the cause of recent climatic change, it is reductive. Framing science as 'settled' strips complexity, nuance and uncertainty from scientific processes and findings.

Uncertainty permeates all branches of science and is not inherently problematic. As Firestein (2012) notes, 'The problem of the unknowable, even the really unknowable, may not be a serious obstacle. The unknowable may itself become a fact.' In climate science, for example, computer climate models are widely used to infer information about past, present and future changes in the climate system. However, a model of the climate system necessarily incorporates many sources of uncertainty[1] (Schmidt et al. 2014; Hargreaves and Annan 2014). Uncertainty arises from an incomplete understanding of climatic processes, from the initial conditions used to start climate model simulations, from the form of mathematical equations used to describe the climate system and from precisely how these equations should be solved computationally. In terms of providing information about the potential impacts of future climate change, uncertainty also arises from the gamut of possibilities of how we as a society will choose to respond to the challenge of climate change. Will we actively cut greenhouse gas emissions or will we choose a business as usual approach? These decision-based possibilities also introduce substantial uncertainties into what climate scientists can say about the future (Knutti and Sedláček 2012).

Although uncertainty is fundamental and unavoidable, the idea of certainty remains attractive in science. The communication of the 'settled science' of climate change conveys this allure. Certain statements can frame interesting and important scientific results. Certain statements can emphasise the value of scientific work. Certain statements can also dispel confusion over the state of scientific knowledge. In practice, scientific studies often sidestep the unknowable in pursuit of elusive certainty, including by using the idea of **significance** to imply some degree of certainty or validity to scientific results. Climate science studies often employ statistical significance testing of correlations, means and trends using **null hypothesis significance testing (NHST)** to convey certainty. The NHST approach is pervasive in many physical sciences and often insisted upon by peer reviewers and journal editors alike.

However, Monash University Professor of Climate Science Neville Nicholls (2001) argues that the NHST approach is flawed for several reasons. He describes a typical null test approach, using a specific study that calculated the correlation between a particular climatic index (the Southern Oscillation Index; McBride and Nicholls 1983) and winter snowfall at a specific location. In this example, the correlation is calculated and a null hypothesis test applied to determine the probability that this correlation could arise if a sample of the same size was drawn from a random population with zero correlation. In cases where the probability (p-value) is less than 0.05, a result is usually regarded as significant.

Nicholls argues that the test is entirely arbitrary and not particularly informative. Who decides what exact p-value conveys adequate significance? Indeed, the 'common belief that the precise quantity 0.05 refers to anything meaningful or interesting is illusory' (in Nicholls 2001, p. 984). Nonetheless, though it is essentially arbitrary, the p-test remains pervasive. In this way, it is commonly used as a falsely certain quantification of what scientists qualitatively already know. Nicholls (2001) notes that it typically 'tells us little of what we need to know and is inherently misleading. We should be less enthusiastic about insisting on its use' (p. 985).

Climate scientists are often prodded to render the indeterminate as known or knowable. For example, scientific answers to societally relevant climate questions would be highly prized. It would be useful to know how heavy rainfall episodes in Sydney will change in 2040 or what changes in hurricane landfall in New York will occur. Such future predictions of climate change at the highly localised scale are most pertinent to population and infrastructure planning and understanding societal vulnerabilities to warming. Hence, such questions are commonly asked of climate scientists. However, the information on climate change most desired by society is often beyond the limits of science. The future is unknown and uncertain. Scientists can make scientific assessments of future climatic change, couched in possible ranges and error bars, and present this information as a basis for decision-making. But scientists cannot even grasp the sources of uncertainty that are encompassed in these assessments; certainty itself is essentially unscientific.

The very concept of a 'settled' climate science is untenable and falsely imbues science with a higher degree of certainty than is possible. Climate scientists certainly have multiple lines of evidence that sufficiently support our understanding of the fundamental behaviour of greenhouse gases in the atmosphere. However, as social scientist Professor John Law (2004) argues, we are never able to 'assert unqualified claims about substances and realities, pin these down, fix them, and make them definite' (p. 28). Rather, scientific statements are inherently qualified and uncertain. This indefiniteness is an aspect of the world that scientists can never circumvent.

In acknowledging the fundamentality of uncertainty, I do not suggest that our understanding of climate change is flawed. Rather, we can understand the world as vague, diffuse, slippery, ephemeral, elusive and indistinct, and *still* appreciate the utility of inherently uncertain scientific knowledge. In the following discussion, I draw out the ungraspable qualities of the world using a case study from climate science—attempts to attribute extreme weather and climate events to a specific cause.

What Caused This?

After an extreme weather or climate event, it doesn't take long for someone to ask 'was this event caused by global warming?' Mike Hulme, Professor of Climate and Culture at King's College London, describes this as the **extreme-weather-blame** question (2014). He defines this question as 'Was this particular weather caused by greenhouse gases emitted from human activities and/or by other human perturbations to the environment?' (p. 2). After another soggy British summer or yet another raging Australian summer heatwave, everyone seems to have an opinion as to the provenance of the event in mind.

In one example, following unseasonably early bushfires just outside Sydney in the spring of 2013, the Australian Prime Minister at the time, Tony Abbott, spoke publicly about the origin of the extreme weather conditions. He described any attempts to link the unusual fires to climate change as 'hogwash,' dismissing links reported in the media between global warming and the fires that razed more than 200 homes and left two people dead (Milman 2013). Abbott's Environment Minister, Greg Hunt, backed his Prime Minister. Minister Hunt disregarded a ministerial briefing from the Australian Bureau of Meteorology on the links between extremes and climate change and instead turned to Wikipedia. He concurred with Abbott that no individual event can be linked to climate change (Davidson 2013). Can scientists determine the causes of individual extreme climate events?

In 2004, researchers at the UK Met Office published a seminal study that attempted—for the first time—to answer the extreme-weather-blame question for a particular extreme climate event from a scientific perspective (Stott et al. 2004). The European summer of 2003 was most likely the hottest since at least A.D. 1500. Conditions were extreme and estimates suggest that excess heat-related deaths topped 15,000 in France alone, and particularly affected elderly and vulnerable people. Peter Stott and his colleagues tackled the extreme-weather-blame question by attempting to estimate by how much human activities may have increased the risk of the occurrence of the 2003 heatwaves. They began by stating 'It is an ill-posed question whether the 2003 heatwaves was caused, in a simple deterministic ways, by a modification of the external influences on climate' (p. 610) and instead proposed an alternative conceptual framework for understanding extreme events.

They started by defining a specific temperature threshold, and using a novel modelling approach combining a suite of global climate model simulations and a methodology borrowed from epidemiologists, they calculated the **fraction of attributable risk** (FAR value) of the event occurring. In the public health arena, a FAR value might provide insight, for example, into the change in risk of developing lung cancer that can be attributed to smoking. In Stott's seminal climate extremes attribution paper, he and co-authors applied a FAR approach to the heatwave and concluded that 'it is very likely (confidence level >90%) that human influence has at least doubled the risk of a heatwave exceeding this threshold magnitude.'

For the first time, a number value was placed on the human influence on an extreme climate event. Stott's paper kicked off an entirely new research endeavour. Since its publication, the FAR approach has been applied to various extreme climate events in different regions. The changed risk of UK floods (Pall et al. 2011), East African drought (Lott et al. 2013), Australian summer heatwaves (Perkins et al. 2014) and Amazonian droughts (Shiogama et al. 2013), amongst other events, have all been quantified. My own work has focused on quantifying the fraction of likelihood of the extreme Australian heat in 2013 that is attributable to human influences (Lewis and Karoly 2013, 2014). Hulme's question 'Was this particular weather caused by greenhouse gases emitted from human activities and/or by other human perturbations to the environment?' has become so widely asked and answered in the scientific literature, it now forms the exclusive basis for an annual special volume of the high-impact *Bulletin of the American Meteorological Society*. In this annual volume, numerous international research groups investigate the causes of particular extreme weather and climate events of the previous year.

The FAR technique represents just one scientific approach for addressing the extreme-weather-blame question. Hulme (2014) summarises three additional broad 'approaches' for understanding the attribution of weather and climate extremes to particular causes:

1. *Simple physical reasoning:* Through simple thermodynamic arguments, we expect more intense precipitation events and more frequent hot temperature extremes due to the enhanced heating from global warming, and the associated increased water-holding capacity of the atmosphere. This approach produces only qualitative statements that indicate whether or not an observed extreme event is consistent with what is known and expected of the anthropogenically enhanced greenhouse effect. Physical reasoning has been used to understand changing spatial and temporal patterns of precipitation (Trenberth et al. 2003).
2. *Statistical analyses:* This approach explores the statistical characteristics of observed meteorological data (i.e. temperature and precipitation),

which are investigated to determine whether a particular weather or climate event falls outside the normal range of what we expect in a 'natural' climate. This approach has been used to investigate extremes such as anomalous European temperatures (Otto et al. 2012) and heavy rainfall in Thailand (Van Oldenborgh et al. 2012).

3. *The end of nature*: This is not a formalised, coherent approach, as such, but a more philosophical viewpoint. This understanding of extremes argues that as there is no doubt that human influences are changing our climate in a pervasive sense, then all-weather events are affected by human influences, because they are occurring in an environment that is warmer and moister than it used to be (Trenberth 2011). There is no longer any such thing as a purely 'natural' weather event.

Regardless of the precise approach employed, the scientific motivations for undertaking extreme event attribution studies are still being untangled. Are these scientific studies aimed at developing new fields and techniques, or improving our capacity for planning for future climate change, or are they communication-oriented, or attempting to highlight the impacts of climate change? Sometime before Stott's ground-breaking paper on the 2003 European heatwaves, Oxford University researcher Professor Myles Allen (2003) published a tantalising commentary in *Nature*. Writing in 2003, as flood waters from the Thames' burst banks lapped at his kitchen door, Allen said that the issue of event attribution is important, as 'it touches on a question that is far closer to many of our hearts than global sustainability or planetary survival—who to sue when the house price falls?' (p. 891). Allen framed the issue, not from a scientific perceptive, but from a legal one. In a warmer world, who will pay for damages? Could current greenhouse gas emitters be held liable for the future impacts of their emissions? The issue of event attribution became an issue of **agency**, with the blame question more precisely framed as 'Can this meteorological event be attributed to human agency as opposed to some other form of agency?' (Hulme 2014, p. 2). If we can attribute extreme weather and climate events to a specific agency, who is to blame?

Although the science of attributing extreme events to a particular cause has evolved rapidly since Allen's (2003) conceptual explanations, understanding causality remains difficult. Causation can be approached from several perspectives, and climate scientists Dr Alexis Hannart and co-authors (2015) employ one specific approach. They suggest that the field of extreme event attribution is hampered by the absence of a formal definition for the word 'cause,' and propose a set of definitions and methodologies to buttress our scientific frameworks. Starting at the point where both 'scientists and philosophers have struggled to define precisely when one event truly causes another and conversely when it does not,' Hannart and colleagues (p. 101) attempt to address causal relationships in extreme weather and climate events.

This search for a formal understanding of causation can be traced back to 1748, when philosopher David Hume (1748, p. 48) suggested that 'We may define a cause to be an object followed by another, where, if the first object had not been, the second never had existed.' Hannart and co-authors (2015) argue that the eighteenth-century understanding of causality remains relevant to science and pose this formally as, where X and Y are events, Y is said to be caused by X if and only if were X not to occur, then Y also would not occur. Hannart uses a simple example to demonstrate causality. We first consider a naive observer, the weather and a barometer, where the observer has no prior knowledge of either meteorology or barometers. After observing that movements of the barometer's needle precede weather changes, the observer infers that the barometer *causes* rainy episodes. If the observer experiments with the barometer and moves the needle, expecting a response in weather, he or she will demonstrate that the barometer does not in fact induce a weather change.

This formalised definition of causality can be extended to the concepts of necessary causality and sufficient causation to form a **probabilistic framework**, which scientists can then apply to extreme climate event attribution:

- The **probability of necessary causation**, PN, is the probability that the event Y would *not* have occurred without the event X, given that both Y and X *did* occur. Hence, PN quantifies how likely it is that X caused Y, although other factors may also be required.
- Sufficient causation, or 'X is a sufficient cause of Y,' means that X always results in event Y occurring, but that Y may also occur for reasons other than X. The **probability of sufficient causation**, PS, is defined as the probability that Y would have occurred in the presence of X, given that Y and X did not occur.

The probability of necessary causation, PN, is useful for formalising an understanding of extreme event attribution. Myles Allen's (2003) courtroom blame analogy is invoked when Hannart and co-authors (2015) argue that the probability of necessary causation matches the reasoning used in lawsuits. In this context, legal responsibility is understood counterfactually, and PN equates to the probability that the damage Y suffered by the plaintiff would not have occurred were it not for the defendant's action X. The accused is declared guilty when it is proven that PN is high enough. The courtroom analogy is extended by Hannart:

> Event attribution thus requires the adversarial debate typical of a lawsuit in order to cautiously balance incriminating versus exonerating evidence. i.e. to evaluate the main cause under scrutiny e.g. anthropogenic **forcings**, as well as each and every possible alternative explanations e.g. Natural forcings or internal variability of the climate system, which may have led to the same outcome... If the resulting PN is high enough, then human responsibility is established and a ruling may in theory follow, as it does in litigation cases. (p. 104)

Using Peter Stott and co-authors' (2004) attribution of the 2003 European heatwave, Hannart formalises Stott's findings as 'the CO_2 emissions are very likely to be a necessary cause, but are virtually certain not a sufficient cause, of the summer of 2003 heatwave. This statement highlights a distinctive feature of unusual events: several necessary causes may often be supported by the data, but rarely a sufficient one' (p. 106).

ADDING A LITTLE BIT OF SPEED

Hannart and colleagues (2015) describe their approach as 'a complete characterization of the causal relationship between X and Y' and that '... these quantities are not nebulous metaphysical notions: the definitions are precise and unambiguously implementable, as long as a fully specific probabilities model of the world is postulated' (p. 103). How does such a complete characterisation sit alongside recognition of the unknowable aspects of the world? This unambiguous rejection of the uncertain is an inflexible understanding of what it means for something to cause something else.

While Hannart's characterisation excludes nebulous metaphysical notions, political theorist Jane Bennett (2009) celebrates them. Bennett proposes a very different understanding of causation to Hannart's mathematical framework, which instead embraces the concept of **vibrant**

materiality. This concept extends the ability to make things happen and to produce effects to nonhuman bodies. These effects and capacities of the nonhuman world create complexities and intractabilities. She describes vitality as the capacity of things—objects or storms—'not only to impede or block the will and designs of humans but also to act as quasi agents or forces with trajectories, propensities or tendencies of their own' (p. viii). Bennett (2009) argues that recognising the vitality of nonhuman matter, and its capacity to produce effects and alter the course of events, provides us with a richer apprehension of the range of powers that transform the world. Bennett describes the idea of **actants** using the ideas of philosopher Bruno Latour, whereby an actant is a source of action, human or nonhuman, or both. An actant as 'something that acts or to which activity is granted by others. It implies no special motivation of human individual actors, nor of humans in general' (Latour 1996, p. 7).

Just a climate scientist Myles Allen's (2003) legal reframing helped communicate attribution approaches, Bennett describes the complex materiality of objects through examples rooted in the legal system. She describes her own experience as juror on an attempted homicide trial. In this criminal case, an item of evidence became an actant. The actant was a Gunpowder Residue Sample. A small glass vial of material had been collected off the accused's hands after the shooting, and provided as evidence as to whether the accused had either fired a gun or been within three feet of a gun firing: this 'composite of glass, skin cells, glue, words, laws, metals, and human emotions had become an actant' (p. 9).

Notably, in contrast to the scientific use of legal analogies for extreme event attribution, Bennett (2009) does not seek to assign blame. She argues that if we are willing to extend our thoughts beyond even an instant in time, a simple 'billiard-ball type' understanding of cause and effect falters. Beyond the most trivial of events (such as occurring during a billiards game), such understandings are inadequate. In recognising both human and nonhuman actants, with complementary or conflicting degrees of capacity, Bennett comes to a more nuanced understanding of causality that is 'more emergent than efficient, more fractal than linear' (p. 33).

In drawing out a more nuanced understanding of causality, Bennett (2009) recalls the work of philosopher Hannah Arendt (1951), who delved into political and historical events to disambiguate the ideas of **cause** and of **origin**. In her work, Arendt eschews causation. Bennet summaries that 'cause is a singular, stable, and masterful initiator of effects, while an origin is a complex, mobile and heteronomous enjoiner of forces' (p. 33). When discussing the causes of political events, such as the rise of totalitarianism in the first half of the twentieth century, we cannot meaningfully understand events in terms of causes, because elements by themselves never cause *anything*. Rather, elements 'become origins of events if and when they crystallise into fixed and definite forms' (p. 34).

This expanded understanding of causation may seem nebulous or too vague to be applied to specific examples. It can, however, be readily demonstrated. In 2013, I was prompted to expand my own thinking of causality when a terrible event occurred right near my university in Melbourne. A tall brick wall around a vacant block blew over on a gusty April day. Two bright young students—siblings—and a visiting postdoctoral researcher were killed in the incident, which is now marked by a lovingly tended memorial.

I couldn't let go of my ruminating; it was impossible to understand. I was also a postdoc at the university and I went by that empty block every day. I even strolled by a few hours before the wall collapsed, and again a few hours afterwards. Everyone wanted to know how something this senseless could possibly happen? Something or someone *must* have been to blame. Perhaps it was the construction company who owned the vacant block? They must have been wholly responsible for the integrity of the brick wall. From a legal perspective, this seems to be the case. In the intervening years, through several independent investigations, the construction company and a company that owned signage on the wall have been charged for breaching health and safety regulations that accelerated the wall collapse (Toscano and Calligeros 2016). Put simply, the 'system failed.'

But several key factors coincided on that tragic day. The unusually strong gusts of wind that kicked up in the mid-afternoon played a role. Advertising hoarding had been bolted onto the old brick wall, making it much taller and weaker than originally intended. It also was a very busy day at the university, with students and staff pouring onto the street following the last day of classes before the extended Easter break. While the construction company was found to have been negligent in their management of the site, can we unambiguously say that they *caused* the tragedy? Was the loss of three young lives a foreseeable outcome of the tall cladding they had erected?

After the deaths, I was appeased by blaming someone, and the implicit reassurance that this could never happen again. While such ready placation has a legal basis, it is physically inadequate as it denies *all* nonhuman agencies. Nonhuman elements in the world—like the wind, the cladding and the sturdy wall, have the capacity to interact with us in complex ways that do not adhere to the human and nonhuman binary. Perhaps the wind, and the wall, and the cladding, and the Easter holidays, and the workers who affixed the sign, and the bricklayers who built the wall, and the mortar, tired and degraded, amongst many other factors, killed the students and young researcher.

What does the complexity contained by human and nonhuman objects mean for climate change and climate science? Attempts to understand the human influence on our complex and dynamic climate system also permit the acknowledgement of powerful nonhuman agents, of vibrant materialities, and of complexities and intractabilities. For example, exceptionally heavy rain fell across Australia between 2010 and 2012 (Bureau of Meteorology 2012). That summer, I spent many soggy months in my flooded backyard, playing totem tennis in the unending rain to relieve that stressed of the tumultuous, final throes of writing my PhD thesis. Over several months, it rained and I wrote, and it rained and I wrote. The rain broke the decade long drought of the preceding decade but was tragically associated with billions of dollars of flood damage, large-scale evacuations and a significant loss of life in Queensland in 2011. There was so much rainfall over the Australian continent during this period that there was a discernible drop in global sea level (Fasullo et al. 2013).

After such large-scale and life-destroying events, we typically turn to the extreme-weather-blame question. However, understanding the physical drivers of this tempestuous period of Australian climate is complicated. The heavy rainfall occurred in association with two unusually strong, back-to-back La Niña events, which are usually associated with wet conditions in eastern Australia (Ganter and Tobin 2013). Furthermore, in a world warmed by human-caused greenhouse gas emissions, we tend to expect an increase in extreme rainfall. Next, the flooding revealed that many houses were built in flood-prone areas. Finally, an ongoing series of inquiries, commissions and inquests suggests that the severity of the floods and loss of lives was exacerbated by catchment-scale mismanagement (Queensland Flood Commissions Inquiry 2012). Were the authorities solely to blame? Or rather did many complex human and nonhuman factors conspire to result in this awful tragedy?

While at first the very concept nonhuman agency may sound farfetched, unfamiliar or irrelevant to a scientific way of thinking about causality, parallel concepts of causation can be useful to science, in addition to mathematical frameworks. The complexities of scientific extreme event attribution are acknowledged by an analogy in the *Bulletin of the American Meteorological Society* 2012 report (Peterson et al. 2013, p. S64), 'Adding just a little bit of speed to your highway commute each month can substantially raise the odds that you'll get hurt some day. But if an accident does occur, the primary cause may not be your speed itself: it could be a wet road or a texting driver.' The climatic analogies to the highway, driver, speeding and texting can be demonstrated in the study of the heavy rainfall in Australia in 2011 and 2012. A suite of studies—including two that I was involved in—have tried to quantify the contributing factors to Australia's extreme rainfall over this period (Christidis et al. 2013; King et al. 2013). The results were ambiguous.

Climate models tend to indicate the human influences increased the odds of having extremely high precipitation in a region, but there are many climatic factors that complicate our understanding of events (remember our wet roads and texting drivers). Complex events, it turns out, are complicated. Equivocal attribution results, such as the causes of Australia's heavy 2011 rainfall, can be viewed simplistically as a scientific 'signal to noise problem.' Rainfall is notoriously complicated—it can be monsoonal one day, and clear the next, and it can be sunny at work and just a few kilometres away, our washing is becoming sadly drenched on the clothesline. This **noise** in the system is large in comparison to the **signal** from anthropogenic warming, and makes attribution studies particularly sensitive to how they are designed. Alternatively, these ambiguous scientific results can be a reminder that, as with my experience reflecting on the wall tragedy, both the human and nonhuman are important.

CAUSES AND ORIGINS

Understanding complex events such as the heavy rainfall or brick wall collapse requires knowledge from a wide repertoire, including literature on the vitality of the nonhuman. This alternative understanding does not suggest that scientific attribution studies are wrong, or that they are not useful. Nor do I suggest the formalised understanding of causality proposed by Hannart and colleagues (2015) and expressed in mathematical terms is incorrect. An alternative understanding does not replace a mathematical understanding, but rather attends to other aspects of the physical world, such as nuances, intractabilities and complexities. I argue that mathematical approaches *alone* do not provide us with a complete understanding of the 'excruciating complexity and intractability' (Bennett 2009) of the human and nonhuman world. In *The End of Nature*, environmentalist Bill McKibben (1990) laments that a contemporary child born today will 'never know a natural summer...Summer is becoming extinct, replaced by something else which will be called "summer"' (p. 55).

Are the human and the natural as dissociable as McKibben argues? I argue that while the physical climate system is now imprinted with human influences, the nonhuman—gusty winds, brick walls, mismanaged dams, formidable La Niña events—also have the ability to act in important, intricate and unavoidable ways. While we usually think of humans as subjects and the nonhuman as objects, such thinking can be reductive. By saying that there is no longer such a thing as 'natural,' McKibben (1990) disregards the importance and capacities of the nonhuman. Attribution studies can improve the skill of climate models and our seasonal forecasting capabilities. They can also provide considered answers about the types of changes in extreme events that we are likely to see under anthropogenic greenhouse gas warming. But these approaches *alone* do not fully equip us to understand disastrous life- and livelihood-threatening extreme weather and climate events.

Can scientists usefully view the human influence on the climate system through the same lens of nuanced causality? If we return to Hulme's (2014) extreme-weather-blame question, we can now see that it is not always as simple as saying that something *can* cause something else. To invoke such reductive explanations of complex events denies a great vitality in the physical world. Furthermore, the concepts of necessary and sufficient causation cannot be a 'complete characterization of the causal relationship between X and Y' (Hannart et al. 2015, p. 103) because these concepts present just one of many ways to understand the world. Of course, there must be more—more than is necessary and more than is sufficient. Indeed, such attempts to develop a singular approach, or canonical understanding of causation, are misplaced because causation is inherently slippery and ungraspable.

Just as an essential unknowability is a fundamental characteristic of the world, 'ungraspability may be an [essential] aspect of agency' (Bennett 2009, p. 36). In terms of understanding agency in extreme climate and weather events, it is particularly problematic to frame the issues in terms of 'blame' and to invoke our well-worn courtroom analogy in addressing the 'extreme-weather-blame question.' This framing necessarily strips the complexity and intractability from our systems. Bennett (2009) argues that fundamentally there is no such juxtaposition of the agent and event, but rather a 'federation of actants...to which the charge of blame will not quite stick. A certain looseness and slipperiness, often unnoticed, also characterises more human-centred notions of agency' (p. 28).

Scientists already acknowledge such slipperiness in our scientific explanations of complex climate events. The *Bulletin of the American Meteorological Society* encapsulated this slipperiness when explaining the complexities of disentangling the actants in extreme climate using their driving metaphor (Peterson et al. 2013). While the additional speed magnifies the odds of a crash, the *Bulletin* remind us of the other possible players in our highway commute story—if 'an accident does occur, the primary cause may not be your speed itself: it could be a wet road or a texting driver' (p. S64). Invoking Arendt's (1951) characterisation of historical events, scientists could shun climatic 'causes' as a singular initiator of events, and rather reclaim such players in the evolution of events and outcomes as origins, complex and mobile. For example, scientists would describe neither human influences nor the strong La Niña as causes of the 2011 heavy rainfall in Australia, but rather discuss them as origins of the eventual tragedy. In this way, extreme-weather-blame question no longer seeks blame as an ultimate goal, but rather, the considered identification of potentially multiple interrelated origins. Evolving causality from a singular characterisation of Xs and Ys to multiple actants and origins does not diminish the value of scientific attribution approaches.

Nonetheless, the value of scientific attribution is difficult to assess while the motivations for such studies remains opaque (*Nature* Editorial 2012). These studies are driven variously by the development of new rational understandings of physical processes and new analytic methods, or to inform climate adaptation strategies, or the possibility of pursuing legal liability for damages (Hulme 2014). These represent disparate, and not necessarily mutually exclusive, motivations for undertaking attribution. These 'mixed and multiple motives at work in the community of scientists' (Hulme 2014, p. 4) are valuable. It is useful to scientists as a group that Hannart et al. (2015) propose a formal, mathematical approach to extremes, *and* that Trenberth (2011) argues that we should seek to prove that human *have not* influenced an event *and* that Hulme (2014) suggests that we must be more introspective about the science of extreme attribution. A single approach cannot encompass a *complete* understanding of our slippery, ungraspable, unknowable world and should not be described as such. While the climate science community should continue to expand the limits of climate modelling at higher temporal and spatial scales, to investigate the contributing factors to ever more complicated extreme weather events, scientists should not overlook the 'excruciating complexity and intractability' (Bennett 2009, p. 4) of the human and nonhuman world, where all are actants and all are potential origins beyond a simple game of billiards.

A modest terminological and philosophical adjustment to origins, not causes, acknowledges the unknowable. The idea of a knowable universe revealed by science permeates public understanding of scientific practice. This misplaced perception is apparent in apocryphal reporting of health sciences—*this* diet is healthy, *this* pill solves everything, *this* exercise leads to weight loss, following rapidly by oops sorry, we meant this other diet/ pill/exercise regime. The known is also a long-held disciplinary value amongst scientists, sitting in tension with our long-held understanding that aspects of our world are simply unknowable. Despite the hard and soft limits on what we can strive to know, science has not willingly embraced uncertainty in our understanding of the practice of science and the communication of its outcomes. Rather, scientists collectively tend to brush off these well-known instances of uncertainty.

I explicated such tensions when I discussed the ideas of settled science, null hypothesis significant testing and Heisenberg's Uncertainty Principle. In doing so, scientists continue to see the world through a Platonic lens in which scientific discoveries uncover essential parts of the world (a **sensible external world**) already in existence. The inherent assumption is that eventually, scientists will be able to reveal everything about the world. This is not a tenable view of the practice of science. Furthermore, climate science continues to crave certainty. Scientific results couched in terms that reflect certainty can help us communicate interesting and important results and emphasise the value of our work within the scientific community.

Nonetheless, scientists cannot genuinely invest in the misplaced ideas of settled science. The unknowable, and all its elusive ripples of uncertainty and slipperiness must be positioned as a *central* disciplinary norm. This does not diminish the value of scientific knowledge. This does not resign science to futility, to the subjective, vague or indefinite. Rather, a relinquishment of the knowable permits scientists a more nuanced and honest understanding of what science can do and what it cannot do.

NOTE

1. The ins and outs of climate models are discussed in detail in *Chapter 3*.

GLOSSARY

Actants   An actant is a source of action, either human or nonhuman, or both. Bruno Latour (1996, p. 7) defines it as 'something that acts or to which activity is granted by others. It implies no special motivation of human individual actors, nor of humans in general.' An actant is neither an object, nor a subject, but an 'intervener.'

Agency   The capacity to act independently.

Anthropogenic   Changes in the environment resulting from human influences, for example, increases in greenhouse gas concentrations from industrialisation.

Extreme weather-blame question   Following Hulme (2014, p. 2) this is 'Was this particular weather caused by greenhouse gases emitted from human activities and/or by other human perturbations to the environment?'

Forcing   Any influence on the climate that originates from outside the climate system itself. For example, changes in solar radiation and greenhouse gas concentrations are forcings.

Fraction of attributable risk (FAR)   The change in risk of an event (such as a heatwave occurring) that can be attributed to particular factors (such as greenhouse gases). This value is widely used in epidemiological studies, for example, probabilistically relating the risk of lung cancer with smoking.

Knowability   The capability of being known, apprehended and understood.

Null hypothesis significance testing (NHST)   A statistical method used for testing whether a select factor has a statistically significant effect on an observation.

Probabilistic   An approach or way of thinking related to probabilities. For example possible scenarios, outcomes or explanations are assessed, with each having a different degree of certainty.

Probability of necessary causation   The probability that the event Y would *not* have occurred without the event X, given that both Y and X *did* occur. This quantifies how likely it is that X caused Y.

Probability of sufficient causation   The probability that an event Y would have occurred in the presence of X, given that Y and X did not occur.

Sensible external world   The idea of a world consisting of objects that exist independently of us, but can be experienced and apprehended through our senses, such as by sight and touch.

Signal/noise   In signal processing, the noise of a system is the unwanted modifications of a signal. More generally, noise is the useless information surrounding useful information (the signal).

Significance   A number used in statistics that expresses the likelihood that a result of an experiment could have arisen purely by chance.

Vibrant materiality   From Bennett (2009, p. viii), referring to the capacity of the nonhuman 'not to impede or block the will and designs of humans but also to act as quasi-agents or forces with trajectories, propensities, or tendencies of their own.'

Uncertainty   I refer to uncertainty broadly, meaning both not knowing and also how well something is known.

Uncertainty principle   A rule of quantum mechanics stating that the more precisely one measure of a particle is made (its position), the less precise another measurement of the same particle (its momentum) will become.

References

Allen M (2003) Liability for climate change. Nature 421:891–892. doi: 10.1038/421891a.

Arendt H (1951) The Origins of Totalitarianism. Schocken, New York.

Bennett J (2009) Vibrant Matter: A Political Ecology of Things. Duke University Press, Durham and London.

Bureau of Meteorology (2012) Australia's wettest two-year period on record; 2010–2011. Special Climate Statement 38. National Climate Centre Bureau of Meteorology, Melbourne.

Christidis N, Stott PA, Karoly DJ, Ciavarella A (2013) An attribution study of the heavy rainfall over Eastern Australia in March 2012. Bulletin of the American Meteorological Society 94:S58–S61.

Cook J, Nuccitelli D, Green SA, et al (2013) Quantifying the consensus on anthropogenic global warming in the scientific literature. Environmental Research Letters 8:024024. doi: 10.1088/1748-9326/8/2/024024.

Davidson H (2013) Greg Hunt uses Wikipedia research to dismiss climate change-bushfires link. In: The Guardianhttps://www.theguardian.com/world/2013/oct/24/greg-hunt-wikipedia-climate-change-bushfires Accessed 12 Aug 2016.

Editorial (2012) Extreme weather. Nature 489:335–336.

Fasullo JT, Boening C, Landerer FW, Steven Nerem R (2013) Australia's unique influence on global sea level in 2010–2011. Geophysical Research Letters. doi: 10.1002/grl.50834.

Firestein S (2012) Ignorance: How it Drives Science. Oxford University Press, New York.

Fitch FB (2014) A logical analysis of some value concepts. The Journal of Symbolic Logic 28:135–142. doi: 10.2307/2271594.

Ganter C, Tobin S (2013) Australia [in "State of the Climate in 2012"]. Bulletin of the American Meteorological Society 94:S196–S198.

Hannart A, Pearl J, Otto FEL, et al (2015) Causal counterfactual theory for the attribution of weather and climate-related events. Bulletin of the American Meteorological Society 150223132524004. doi: 10.1175/BAMS-D-14-00034.1.

Hargreaves JC, Annan JD (2014) Can we trust climate models?. WIREs Clim Change. doi: 10.1002/wcc.288.

Hulme M (2014) Attributing weather extremes to "climate change": A review. Progress in Physical Geography. doi: 10.1177/0309133314538644.

Hume D (1748) Enquiry Concerning Human Understanding. Reprinted Open Court Press 1958, LaSalle, USA.

IPCC (2013) Climate change 2013: The physical science basis. Contribution of Working Group I to the Fifth Assessment Report of the Intergovernmental Panel on Climate Change [Stocker TF, Qin D, Plattner G-K, Tignor M, Allen SK, Boschung J, Nauels A, Xia Y, Bex V and Midgley PM (eds.)]. Cambridge University Press, Cambridge, UK and New York, NY. 1535 pp. doi:10.1017/CBO9781107415324.

King AD, Lewis SC, Perkins SE, et al (2013) Limited evidence of anthropogenic influence on the 2011–12 extreme rainfall over southeast Australia. Bulletin of the American Meteorological Society 94:S55–S58.

Knutti R, Sedláček J (2012) Robustness and uncertainties in the new CMIP5 climate model projections. Nature Climate Change 3:369–373. doi: 10.1038/nclimate1716.

Latour B (1996) On actor-network theory: A few clarifications. Soziale Welt. doi: 10.2307/40878163.

Law J (2004) After method: Mess in social science research. Routledge, London and New York.

Lewis SC, Karoly DJ (2013) Anthropogenic contributions to Australia's record summer temperatures of 2013. Geophysical Research Letters. doi: 10.1002/grl.50673.

Lewis SC, Karoly DJ (2014) The role of anthropogenic forcing in the record 2013 Australia-wide annual and spring temperatures [in "Explaining Extremes of 2013 from a Climate Perspective"]. Bulletin of the American Meteorological Society 95:S31–S34.

Lott FC, Christidis N, Stott PA (2013) Can the 2011 East African drought be attributed to human-induced climate change?. Geophysical Research Letters 40:1177–1181. doi: 10.1002/grl.50235.

McBride JL, Nicholls N (1983) Seasonal relationships between Australian rainfall and the Southern Oscillation. Monthly Weather Review 111:1998–2004.doi: 10.1175/1520-0493(1983)111<1998: SRBARA>2.0.CO;2.

McKibben W (1990) The End of Nature. Viking Books, London.

Milman O (2013) Climate change linked to bushfire risk says Environment Department website. In: The Guardian. https://www.theguardian.com/science/2013/oct/25/climate-change-linked-to-bushfire-risk-says-environment-department-website Accessed 12 Aug 2016.

Nicholls N (2001) The insignificance of significance testing. Bulletin of the American Meteorological Society 82:981–986.

Otto FEL, Massey N, Van Oldenborgh GJ, et al (2012) Reconciling two approaches to attribution of the 2010 Russian heat wave. Geophysical Research Letters. doi: 10.1029/2011GL050422.

Pall P, Aina T, Stone DA, et al (2011) Anthropogenic greenhouse gas contribution to flood risk in England and Wales in autumn 2000. Nature 470:382–385. doi: 10.1038/nature09762.

Perkins SE, Lewis SC, King AD, Alexander LV (2014) Increased simulated risk of the hot Australian summer of 2012–2013 due to anthropogenic activity as measured by heatwave frequency and intensity [in "Explaining Extremes of 2013 from a Climate Perspective"]. Bulletin of the American Meteorological Society 95:S34–S37.

Peterson TC, Hoerling MP, Stott PA (2013) Explaining extreme events of 2012 from a climate perspective. Bulletin of the American Meteorological Society 94: S1–S74. doi: 10.1175/BAMS-D-13-00085.1.

Queensland Flood Commissions Inquiry (2012) Queensland Floods Commission of Inquiry: Final Report.

Schmidt GA, Annan JD, Bartlein PJ, et al (2014) Using palaeo-climate comparisons to constrain future projections in CMIP5. Climate of the Past 10:221–250. doi: 10.5194/cp-10-221-2014.

Shiogama H, Watanabe M, Imada Y, et al (2013) An event attribution of the 2010 drought in the South Amazon region using the MIROC5 model. Atmospheric Science Letters. doi: 10.1002/asl2.435.

Stott PA, Stone DA, Allen MR (2004) Human contribution to the European heatwave of 2003. Nature 432:610–614. doi: 10.1038/nature03089.

Toscano N, Calligeros M (2016) Wall collapse: How the system failed. In: The Age. http://www.theage.com.au/victoria/wall-collapses-at-north-melbourne-construction-site-20160420-goakg3.html Accessed 10 Aug 2016.

Trenberth KE (2011) Attribution of climate variations and trends to human influences and natural variability. WIREs Clim Change 2:925–930. doi: 10.1002/wcc.142.

Trenberth KE, Dai A, Rasmussen RM (2003) The changing character of precipitation. Bulletin of the American Meteorological Society 1205–1217. doi: 10.1175/BAMS-84-9-1205.

Van Oldenborgh GJ, Van Urk A, Allen M (2012) The absence of a role of climate change in the 2011 Thailand floods [in "Explaining Extremes of 2011 from a Climate Perspective"]. Bulletin of the American Meteorological Society 1047–1067. doi: 10.1175/BAMS-D-11-00021.1.

CHAPTER 3

The Pseudo in Our Science

Abstract Lewis explores the ways in which scientific knowledge claims are legitimised through assessing their falsifiability. Beginning with an exploration of the construction and use of climate models, Lewis questions whether climate models are falsifiable. From a scientific perspective, knowledge that is not falsifiable is considered pseudo-science. By exploring the model tools used in climate science, Lewis problematises this traditional binary between science and pseudo-science. In response, Lewis poses a new way of evaluating knowledge claims, which are alternatively viewed on a spectrum and assessed by their usefulness. This spectrum of knowledge is applied to various climate sceptic claims around temperature records and wind power to demonstrate the value of Lewis's reappraisal of knowledge claims.

Keywords Climate models · Falsifiability · Pseudo-science · Spectrum of knowledge · Legitimacy

The idea of **legitimacy** forms the second entry within the assemblage of scientific concepts, the key set of ideas and principles that characterise science. Science invests in the ideas of **falsifiability** and reproducibility as a means of demonstrating the value and legitimacy of scientific knowledge. Falsifiability is the idea that an assertion can be shown to be false by an experiment or an observation, and is central to scientific distinctions between 'true science' and '**pseudo-science.**' *All* knowledge claims that

© The Author(s) 2017
S.C. Lewis, *A Changing Climate for Science*,
DOI 10.1007/978-3-319-54265-2_3

are not falsifiable are found to be scientifically lacking. This chapter problematises this knowledge binary by exploring the legitimacy of climate science knowledge derived from climate models. It probes a central question—are climate models falsifiable?

This chapter challenges the utility of the long-standing scientific ideal of falsifiability through discussion of how climate models are constructed and used by climate scientists. I provide examples from my own experience of using climate models to research how extreme weather and climate events are linked to global warming. Using these examples, I argue that the science/pseudo-science binary, as determined by falsifiability, is limiting and excludes swathes of contemporary science. In response to the limitations of a scientific binary determined by falsifiability, I propose a spectrum of scientific knowledge that is assessed by usefulness.

KNOWLEDGE FROM CLIMATE MODELS

In the field of climate science, computer climate models are typically used to address various research questions. My own work generates **simulations** of the climate using computer models, compares theses model results to **observations** of climate, and then uses the model data to understand changes in the climate system (Lewis and Karoly 2014). More specifically, using global climate models, I research how recent extreme weather and climate events are linked to global warming. To reveal the link between extreme events and global warming, my research calculates the likelihood[1]. of particular extreme weather and climate events occurring in alternative scenarios of our world's climate history. This technique determines the probability of extreme climate events (such as heatwaves) occurring in different climate model simulations.

In the first **experiment**, the climate model is run with increasing concentrations of human-caused greenhouse gases, such as carbon dioxide (CO_2) (Stott et al. 2012). The likelihood of a heatwave occurring is calculated for this particular scenario, and then compared to the likelihood of the heatwave happening in a parallel set of climate model experiments (see Fig. 3.1), in which greenhouse gas concentrations are set to much lower levels, such as those that occurred before industrialisation began in around 1850. This approach attempts to create a hypothetical world that might have existed without any human influence on the climate from industrialisation. The difference in the event risk values for these two sets of climate model experiments gives us a quantitative estimate of link

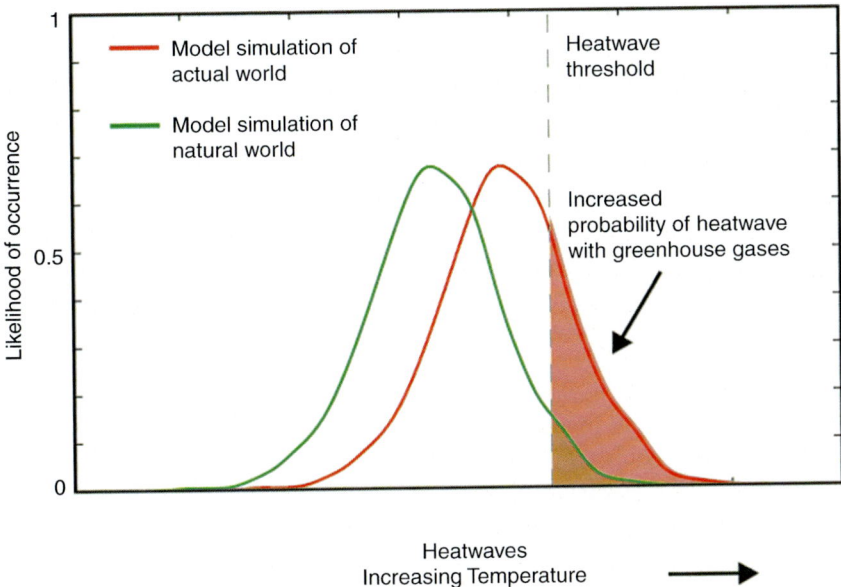

Fig 3.1 Climate model experiments can show how the influence of human-caused greenhouse gases affects the probability of an extreme climate event occurring. Climate model simulations are made using all known climate forcings, and the likelihood of an extreme event occurring, such as a heatwave, is calculated (red line). This is compared to the likelihood of the same heatwave occurring in climate models run without any human influences from greenhouse gases (green). Any difference in the probability of a heatwave in the models reveals the linkage between global warming and this particular extreme event.

between an extreme event and global warming. In *Chapter 2,* we saw that the devastating record-breaking European summer of 2003 was threefold more likely to occur due to human influences (Stott et al. 2004).

Explorations of human influences on the risk of extreme events occurring are just one example of research that utilises climate model simulations. Overall, much of our collective understanding of the physical climate system is underpinned by results derived from climate models. These tools are used broadly by climate scientists to interrogate many aspects of the climate system beyond the human influence on extreme climate events, including into the nature of interactions between the various, interconnected components of the Earth. That is, climate models are instrumental to climate scientists in generating scientific knowledge.

The knowledge derived from climate models is routinely offered to policymakers, and used to impel action on climate change (Weart 2010). For example, 2015 ended with anxious international talks in Paris, which were the latest in a long series of meetings of the United Nations Framework Convention on Climate Change (UNFCCC). The hard-fought Paris Agreement brought together 195 countries in a treaty stipulating the mitigation of greenhouse gases emissions beginning in 2020. The convention aims to hold 'the increase in the global average temperature to well below 2 °C above pre-industrial levels and to pursue efforts to limit the temperature increase to 1.5 °C above pre-industrial levels, recognizing that this would significantly reduce the risks and impacts of climate change' (UNFCCC 2016, p. 3). This was widely described as an ambitious and historic plan to tackle climate change by accelerating our collective response. The hope-filled Paris goals, and their political and economic implications, were founded on the outcomes of climate models, amongst other data sources. Given their key function in sociopolitical decision-making, I explore how climate models work and the knowledge they provide scientists.

The development of a computer model of the climate system is an onerous task, riddled with both scientific and conceptual challenges. First, the construction of a climate model is technically difficult, requiring a solid theoretical understanding of many complex physical processes. Second, a thorough conceptual understanding of climate models is useful for appraising their contribution to scientific knowledge. Let's first focus on the conceptual challenges of modelling. By the orthodox understanding of science and its ways of knowing that were first discussed in *Chapter 1*, scientists primarily obtain truth through the scientific method. This method is well demonstrated by the example of Edward Jenner and his highly successful smallpox vaccination. The scientific process of inquiry requires observation, hypothesis, further observation and finally, the generation of a theory. This traditional (though simplified) understanding of science assigns a linear position to theory and fact: facts appear from theories. The process of inquiry is apparent and describable, when science has established a solid, robust hypothesis, facts emerge and scientific truth ensues. The orthodox relationship between facts and theory prompts questions about the place of climate models in scientific practice. How well do climate models as the modern tools of climate scientists align with this process of discovery?

In the public domain, climate change is intensely debated. I have contributed numerous articles to an online academic news site, *The Conversation*, where public interest in climate change research is high, and climate change research is widely criticised. In this online space, my scientific results and the techniques I use to generate scientific knowledge are publicly scrutinised. The long, meandering comment threads of news websites reveal a wide range of views that funnel in specifically on the use of climate models by climate scientists. The use of climate models has become increasingly politicised and the public response to model-based scientific studies is often characterised by a barrage of heated exchanges. Dedicated blogs such as *Watt's up with that?* focus intently on discrediting models of climate change. Climate sceptics doubt the reliability of climate scientists, scientific pronouncements, *and* the very use of climate models. According to sceptics, climate models are flawed, unreliable and simply *cannot* be trusted.

Climate scientists can respond in several ways. We can choose to dismiss sceptics' criticisms as merely aspersions cast by those with dubious motivations. Alternatively, climate scientists can choose to recognise these questions around climate models as having potentially legitimate foundations, regardless of their motivations. As a community, scientists have set our *own* benchmarks for the legitimacy of scientific knowledge, resting firmly on the process of the scientific method and the key idea of falsifiability as a separator of true science and pseudo-science. The empirical sciences tend to be on produce measurable results Sound science is data rich, built on observations, experiment and theory; theory brings about calculations, and experiments (Guillemot 2010). What about climate science? As detailed in *Chapter 2*, this branch of science is imbued with an inherent uncertainty and complexity that permeates our key instruments, climate models. What 'kind' of science is climate science? Is it true science or pseudo-science?

TOOLS OF DISCOVERY

Let's begin digging into the nuts and bolts—the technical challenges—of climate models. Climate models are comprised of a bundle of mathematical equations that are based on **fundamental natural laws**, such as laws around the conservation of energy, mass and momentum. Global climate models

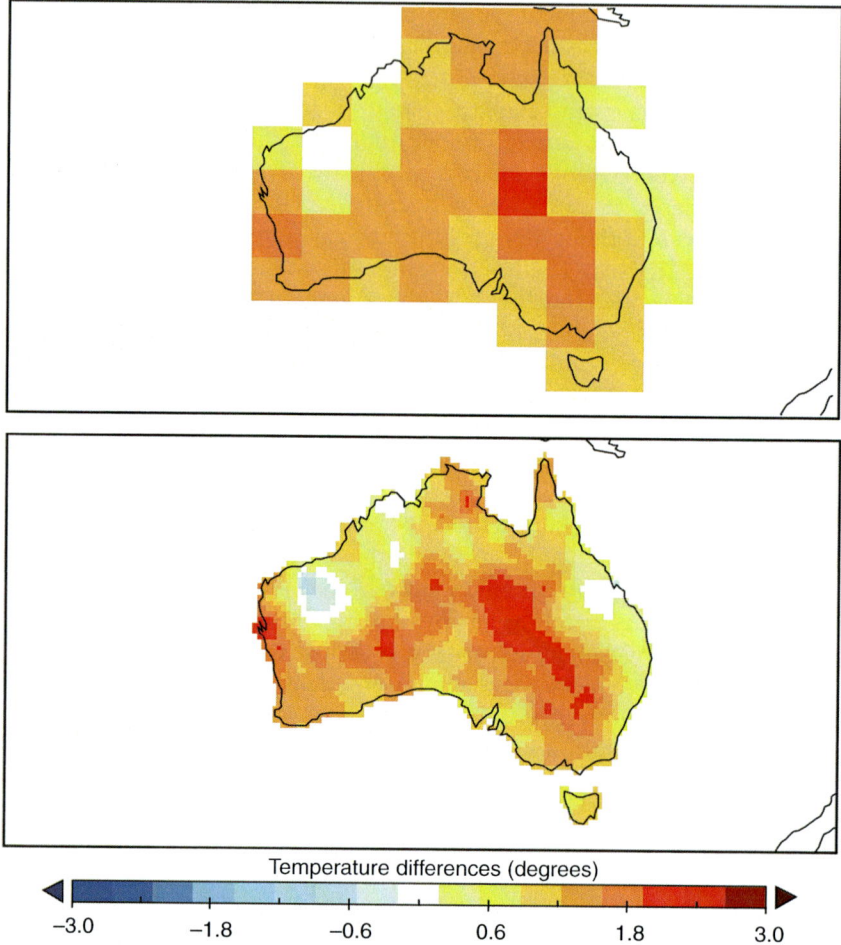

Fig. 3.2 Example of breaking the world up into gridboxes for Australian temperatures. The upper panel shows a coarse resolution model grid (4° latitude by 5° longitude), while the lower panel shows a higher resolution model grid (0.5° latitude by 0.5° longitude). The higher resolution grid tends to produce more realistic model results, but is more computationally demanding to undertake.

typically divide up our spherical world and force it onto a flat, three-dimensional grid made up of latitude, longitude and height coordinates. In each model gridbox, the values of numerous climatic variables, such as surface temperature, pressure, humidity and rainfall are calculated over time using a computer (see Fig. 3.2). Climate scientist Dr Gavin Schmidt (2007) uses three broad categories to describe the physics contained in a climate model:

1. The first category encapsulates fundamental physical principles, such as the conservation of energy and momentum.
2. The second category describes physics that are theoretically well known, but must be **approximated**. More specifically, this process of approximation is through the **discretisation** of continuous equations. For example, the transfer of radiation through the atmosphere is approximated through this process of breaking equations up into discrete parts that can be solved on a computer.
3. The third category contains physics known **empirically**, such as formulas for evaporation as a function of wind speed and humidity.

The process of constructing a climate model requires several steps to incorporate these categories of physics into a single working model. First, the climate system's fundamental physical laws need to be expressed in mathematical terms that can be solved using a computer. This is a challenge of Category 2—as continuous mathematical equations cannot be solved precisely and simply, they need to be transformed into approximated counterparts that can be implemented into our three-dimensional model Earth. Ideally, these estimate mathematical solutions are approximately correct, although the process of discretising of these continuous equations necessarily introduces errors into model calculations. In addition, these errors are impossible to assess. As scientists do not have the continuous solutions to the equations available for comparison, these errors cannot even be accurately estimated (Mueller 2010).

Next, scientists use **parameterisations** to deal with the physics of the climate system that scientists cannot describe explicitly (Category 3). These parameterisations help capture physical processes that are highly complex (e.g. biochemical processes within vegetation) or processes that occur at small spatial and/or temporal scales (e.g. cloud processes). For example, the formation, growth and precipitation of clouds occur on micro-scales of less than 1 mm that cannot be represented directly using comparatively coarse model gridbox scales. The use of parameterisations is unavoidable, and ideally does not present any significant hurdles to modelling the climate system with accuracy. Model developers aim for 'good empiricism' (Petersen 2000), whereby parameterisations are based on what scientists observe, they are robust under different conditions and then left unchanged as the model is tested as a whole, rather than tweaked on the go. Developers also aim to create a model in which the behaviour of the climate system is not dependent on the parameterisations used. However, regardless of their 'goodness,' parameterisations remain only representations of key climate processes.

This climate model—this tangle of physics encapsulated by mathematical expressions approximated using sophisticated numerical methods (discretisation and parameterisations)—is then turned into code that can be executed on a computer, with the execution of this complex code typically performed on a supercomputer. These approximations demonstrate that climate models are *not* fully theoretically based, but instead climate models incorporate at least some degree of arbitrariness. Overall, a climate model integrates many sources of uncertainty, including from these decisions about model parameterisations and about the best form for our mathematical equations and how they should be solved computationally. In addition, uncertainties in climate modelling arise from the initial conditions used to start a model simulation going, the boundary conditions (such as greenhouse gases) used to drive a simulation through sequential time steps, and from imperfect observations of the climate system.

Climate scientists use several approaches to try to understand these uncertainties in climate models. For example, many studies employ an **ensemble** approach to estimate uncertainty in a particular aspect of the climate system (Parker 2013). An ensemble approach might use the results from several different climate models to assess uncertainty in the climate system's **internal variability** (i.e. natural variations) or uncertainty of the response of the climate system to **external forcings** (i.e. drivers of change such as greenhouse gases). In this way, climate models are used collectively as complementary resources (Parker 2006). Generally, when different models within an ensemble agree about a particular aspect of future climate change, findings are described as robust and are often considered more likely to be true (Parker 2011).

An ensemble can be defined in several ways, including as a collection of model simulations performed using multiple climate models, or alternatively as multiple experiments performed by a single climate model with slightly different model parameters or different initial conditions

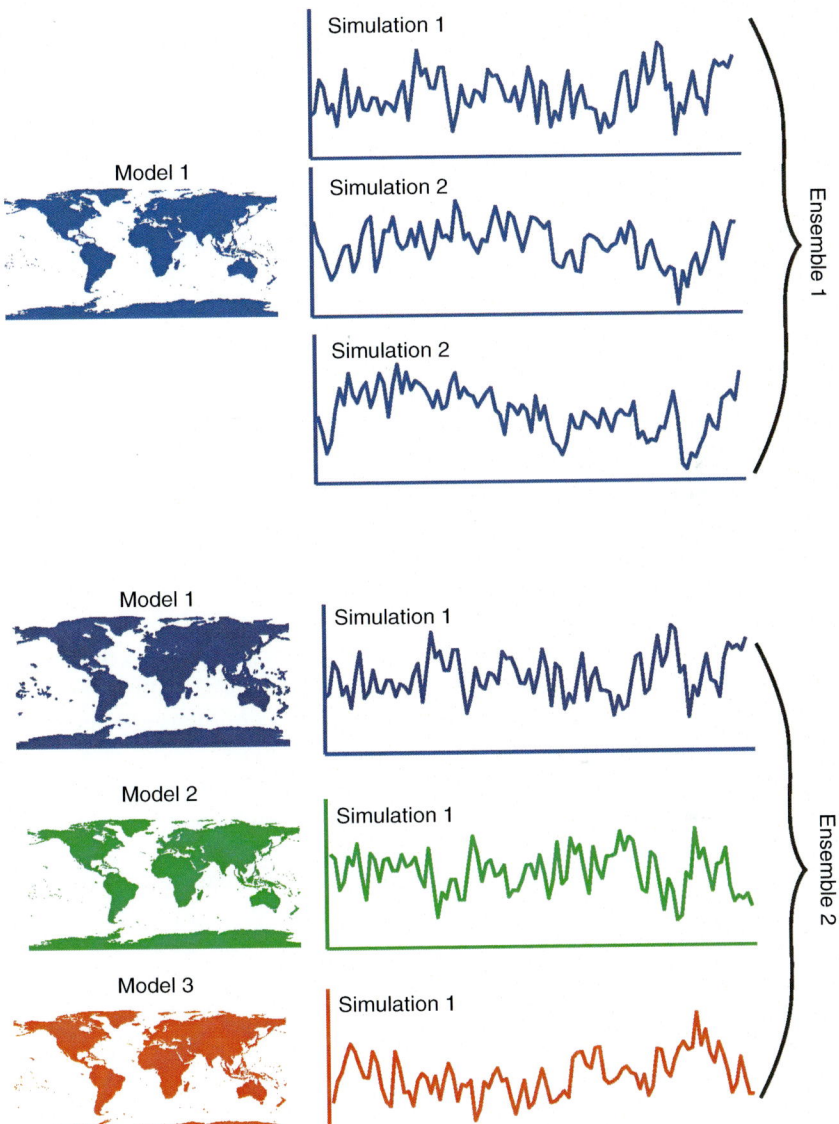

Fig. 3.3 An ensemble of climate models can be defined in several ways. The top panel shows an ensemble formed by multiple simulations of climate using the same climate model run with different starting conditions. The bottom panel shows an ensemble formed by bringing together simulations of climate from different climate models.

(see Fig. 3.3 for further explanation). Ensemble approaches typically interpret the spread (or range) of results across a climate model ensemble from a probabilistic perspective. In this interpretation, given a large enough set of models, the 'truth' is seen as lying in the middle of the ensemble of model simulations.

However, an ensemble of model simulations can be understood in several ways (Sanderson and Knutti 2012; Haughton et al. 2014). Climate scientists can consider each model (or simulation) as an approximation of the true system (the real world), with some additional degree of random error. Any discrepancies between a model and the real world are then just 'noise' in the system. Conversely, scientists can consider the true climate system as theoretically indistinguishable from a model simulation. From this view, each model simulation and the real climate system are grouped together, and can be considered exchangeable with another.

Before climate scientists leap in and use the results from a climate model, we assess how **skilful** climate models are at capturing key features of the observed climate system. An evaluation of model skill occurs on two levels (Schmidt 2007). First, models are evaluated on small scales, where the specifics of the chosen parameterisations are tested. For example, various aspects of climate are tested in models where different technical decisions about clouds and precipitation have been made, which are called convective parameterisations (Knutson and Tuleya 2004; Lopez et al. 2009).

Next, models are evaluated at scales where large phenomena emerge from the complex interactions of small-scale physical systems. These **emergent properties** are a surprising outcome of climate models. Schmidt (2007) notes, for example, that scientists do not include a formula in climate models that explicitly describes the Intertropical Convergence Zone (ITCZ). Instead, the ITCZ—the thick band of convective cloud that controls tropical rainfall—results from the complexity of the modelled climate system. This complexity integrates, for example, the processes of the seasonal cycle of incoming solar radiation, moist convection and the rotation of the Earth. The successful emergence of these large-scale phenomena within climate models provides a useful test bed for assessing the skill of climate models in capturing key components of the observed system.

When model simulations are evaluated against observations, scientists hope that our models produce physically realistic ('right') answers for physically plausible ('right') reasons. Parker (2006) notes that 'a model may be praised or faulted either on the basis of how well its assumptions mesh with existing background knowledge about the climate system or on the basis of the perceived quality of its simulations' (p. 357). As part of the Intergovernmental Panel on Climate Change (IPCC) reports, climate models are systematically compared with observations. The IPCC has concluded in successive iterations that climate models do capture key features of past and present climatic change with skill (IPCC 2007, 2013). These methodical evaluations demonstrate that climate models are skilful in reproducing important aspects of notable past climate events, such as the Last Glacial Maximum and the Little Ice Age, and the historical increase in global average surface air temperature in response to anthropogenic greenhouse gases.

While climate models demonstrate skill in capturing many of these observed climate processes, both technical and conceptual challenges remain (Mueller 2010). It is impossible to account for and incorporate all the processes within the observed climate system into a model. It is impossible to accurately include processes occurring on all spatial and temporal scales. The necessity of using an ensemble of climate models demonstrates these remaining scientific challenges. Ideally, climate scientists would simply select a single best model to use thereafter for understanding every aspect of the climate system. However, when modelled temperature or precipitation is compared to observed climate datasets, no single model consistently performs best (Parker 2006). Some models perform better for some climatic variables, while others perform better against some observational datasets. Since models differ in many different ways, climate scientists are left with a plurality of climate models.

The practice of climate science without climate models is difficult to imagine. We have only one climate system. It has only one climate history. It does not permit us to conduct controlled experimentation. Such limitations of real-world observations and experimentation are succinctly put by Knutson and Tuleya (2005, p. 5183): 'if we had observations of the future, we obviously would trust them more than models, but unfortunately observations of the future are not available at this time.' As such, scientists need climate models to represent our intricate world. These models are powerful, flexible tools that can be used to address many questions about the climate system, which are impossible to interrogate using the 'real' system alone. However, the use of knowledge from climate models throws forth many questions. How do models fit into a traditional understanding of the production of scientific knowledge? How can climate scientists reconcile the uncertainties, the arbitrariness of parameterisations, and our difficulties in evaluation with the knowledge claims made by climate models?

SCIENCE OR PSEUDO-SCIENCE?

In *Chapter 1*, I detailed Popper's (1963) idea of falsifiability, whereby an assertion must be able to be shown false by an experiment or an observation. Popper states, 'In so far as a scientific statement speaks about reality, it must be falsifiable; and in so far as it is not falsifiable, it does not speak about reality.' Knowledge claims that are not falsifiable are found to be scientifically lacking, and these claims are instead classified as pseudo-science. Here I use the term pseudo-science from a firmly *scientific* perspective. A popular definition of pseudo-science describes pursuits such as astrology, homeopathy, phrenology, and a collection of various unruly self-described experts in tinfoil hats. While a scientific understanding of pseudo-science also encompasses these activities, it is both far broader and far more clearly distinguished than the popular usage. Using tests of falsifiability, scientists establish both a benchmark of scientific legitimacy and define a dichotomy. On one side of this dichotomy, scientists place falsifiable science and on the other side, we must place pseudo-science.

What about climate models? There is a high level of public distrust in climate models, with sceptics claiming that climate models themselves are disreputable and cannot be trusted. Are they correct? Which side of the science/pseudo-science binary do climate scientists inhabit? In order to tackle these questions, I will start by posing a question—is knowledge derived from climate models falsifiable? Even this simplified starting point is difficult to untangle. In asking whether climate models are falsifiable, climate scientists might choose to focus on attempting to understand how well climate models simulate observations of global warming. A test of falsifiability requires a model test or climate observation that shows global warming caused by increased anthropogenic greenhouse gases is untrue. So what kind of behaviour would climate scientists have to observe in the climate system that would be inconsistent with climate model based predictions of warming, to the degree that scientists would categorise models as essentially wrong?

Climate scientists Dr Hargreaves and Dr Annan (2014) argue that climate model predictions are, in actuality, trivially falsifiable. They identify predictions about global annual mean temperature rises over the twenty-first century in response to increasing greenhouse gas emissions as the most important result from climate model-based studies. Models project a rise in temperature of around 2–4°C over this period, so in order to falsify the hypothesis represented by climate model projections, we simply wait some 100 years or so and compare current model based predictions with the ensuing observed climate change. However, they further make the point that in *practical* terms, climate models are unfalsifiable. That is, we don't want to have to wait 100 years to find out if our model-based predictions are correct. Instead society wants to know as a matter of urgency whether we can interpret climate model predictions as truthful and act accordingly.

The *My View on Climate Change* blog (2013), collated by atmospheric scientist Dr Bart Verheggen, extends this approach and offers a suite of tests of falsifiability of the human impact on the climate system. This list includes (with my emphasis added in italics):

- A drop in global temperatures for *some period of time* to the level of *50 years ago or longer, with no clear cause*
- A drop in global sea level *for some period of time*
- The discovery that climate forcings in the past *were much larger*, or temperature changes much *smaller*, than science *thinks*
- *Major* errors in equipment in satellites, measuring outgoing long-wave radiation

The emphasis I have added here is key to untangling the challenge of model falsifiability. How robust are these tests of falsifiability, given they depend so intimately on the highly subjective degree and timescales of change? I argue that these examples, together with Hargreaves and Annan's (2014) 100-year model versus observations comparison, are not necessarily valid tests of falsifiability.

First, the actual state of the climate we experience represents just one possible outcome of chaotic, natural climate variability. The significance of natural climate variability was recently demonstrated in the rancour about the hiatus in the rate of global warming (Hawkins et al. 2014). After rising rapidly in the 1990s, global average temperature increases at the Earth's surface have slowed since 1998. The observed change in the rate of warming at the surface ignited all sorts of sceptical claims that global warming had stopped and in response, all sorts of scientific studies aimed at explaining the slowdown (Kosaka and Xie 2013; England et al. 2014; Huber and Knutti 2014; Roberts et al. 2015). Regardless of this concerted scientific effort, sceptics claim that climate models failed to predict the hiatus and hence have systematically failed as scientific tools (see discussion and comments in Tollefson 2016).

In this real-world example, falsifiability is again reduced to a matter of degrees. If scientists cannot *a priori* propose an objective test, then the falsifiability of models may be questionable. In addition, the implications of these tests are unclear. If *one* of these tests fails, have scientists falsified climate models, their predictions of climate change, or the theory of anthropogenic global warming? More broadly, if we play the role of an impartial assessor of falsifiability, can we objectively distinguish between the sceptic claims that models have been proved false, and scientific claims that the test of falsifiability posed by the hiatus is not sufficient as this is not long enough/not substantial enough to separate from natural climatic variability? And what about the implications for the scientific status of climate modelling and climate science if falsifiability eludes us?

The concept of falsifiability is handy, and provides scientists with a ready-made test of the legitimacy of scientific knowledge and the invalidity of other knowledge claims. For example, a recent attack on the credibility of Australian climate scientists demonstrates the utility of such a clear division of knowledge. In 2014, Australian observed temperature records underwent intense scrutiny, driven by claims that scientists at the Australian Bureau of Meteorology (BoM) had falsified measurements to exaggerate warming trends over the twentieth and twenty-first centuries.

Sceptic claims were nourished by a series of high-profile articles in *The Australian* newspaper, which aired perceived grievances of fraudulent behaviour by Australian government scientists (Lloyd 2014). Such attacks can be galling. Australian temperature records are the most highly scrutinised and rigorously controlled in international science, having undergone expert peer review (Trewin 2012). Rushing to defend the Bureau's integrity, sceptic criticisms were discredited by labelling them as 'pseudoscience.' For example, responding in a rival newspaper, Monash University scientist Associate Professor Michael Brown (2014) described the attacks as embracing pseudo-scientific tactics.

This 2014 exchange around the legitimacy of Australia's temperature records reveals the limitations of the science/pseudo-science dichotomy, clearly delineated by tests of falsifiability. While the substance of the attacks on Australian climate scientists were without foundation, this discourse sets an unforgiving standard. If scientists denigrate other knowledge claims as pseudo-scientific, then surely science is obligated to be strictly 'scientific' at all times? Recall that I am discussing pseudo-science from a scientific perspective, meaning knowledge claims that are unfalsifiable, rather than using the popular terminology. Hence, when climate scientists discuss climate models simulations as 'consistent with' or 'inconsistent with' observed climate phenomena, we are viewing our disciplines through this science/pseudo-science binary. If scientists embrace this knowledge dichotomy, then science must pursue knowledge through strictly 'scientific' means at all times; any science that deviates from this approach is, by definition, not science.

Such intransigent definitions rapidly become restrictive, and are readily problematised. Several areas of research can be easily defined as 'science' using this orthodox viewpoint: analytical chemistry, plant genetics, geology and so forth. In these largely empirical disciplines, a hypothesis can be posed, an experiment carefully conducted and a theory falsified. However, other branches of scientific research are, contradictorily and confusingly, pseudo-science. I have described the impediments of classifying climate science as 'true science' and I argue that it is equally unclear if elements of astrophysics, theoretical physics and probability theory are true science or if they are pseudo-science. As climate science and other dubiously falsifiable branches of science are evidently science, questions of falsifiability are clearly ill-posed and ill-applied.

A Spectrum of Knowledge

In response to these limitations, I propose an alternative approach to viewing knowledge. But first, I begin by introducing a different way of viewing **problems**. Philosophers Gilles Deleuze and Félix Guattari discuss problems in terms of usefulness (see discussion in Gaffney 2010). In this sense, 'problems' do not refer conventional, ordinary problems to be solved, but rather problems are an essential part of a creative process. Academics are not attempting to 'solve' problems but rather than to pose them; problems spur us into being. From this alternative viewpoint, we can assess knowledge claims very differently from the pseudo-science approach. In a scientific context, we know that all scientific knowledge is essentially 'wrong.' Some scientific knowledge is wrong and so it is discarded, while some science is thought now to be right but will one day be shown to be wrong and will be discarded, and some science is wrong and but embraced anyway.

Hence, it is not necessarily valuable to talk about 'right' or 'wrong.' Rather it is valuable to view knowledge as a useful because of what it does. Problems—in the Deleuze and Guattari sense—are useful because of their capacity to incite a response, or to catalyse an action. In a classic and well-worn example, scientists still invest heavily in the teaching and use of Newtonian mechanics, although it has been exhaustively demonstrated as limited. Regardless of 'right' or 'wrong,' analysing the world from a Newtonian perspective remains *useful*. If instead scientists focus on utility, we longer need to invest in evaluating whether knowledge claims are inherently 'right' or 'wrong.' Furthermore, a focus on utility presents a spectrum of knowledge claims evaluated by usefulness, rather than a simplistic science/pseudo-science binary, as defined by scientists. Within such a spectrum, the legitimacy of falsifiable science, non-falsifiable science and popular pseudo-science can be evaluated, without unwittingly creating contradictory pseudo-scientific disciplines of science, such as climate science.

How would such a spectrum of knowledge be used and usefulness assessed? For a start, what we popularly understand as pseudo-science (our homeopathy, astrology and foil-hatted prophets) can be broadly disregarded as lacking in use. Why so? These claims by pseudo-scientists will largely be discarded using our revised knowledge spectrum, but not simply because of ideas about falsificationism as necessary for distinguishing knowledge. Now speaking about pseudo-science from a popular perspective, such claims can be useful to some degree. Most obviously, it can be useful for climate sceptics, for example, to cast shadows of doubt on scientific evidence of climate change to serve their own ideologies and agendas (Oreskes and Conway 2010).

The utility and rewards of such tactics are demonstrated in the almost commonplace Australian parliamentary inquiries into temperature data that effectively distract scientists from their core work. More broadly, it can be useful to understand why people reject peer reviewed data in favour of falsely manipulated temperature datasets that mask the clear signal of global warming. It can be useful to understand why wind turbines incite loathing, or why the value of vaccinations is contested or why some believe that their fate is written in the stars.

My own dabbling in tarot shows the utility in popular pseudo-science. In mid-2011, I was metaphorically lost. Months before, I had submitted my PhD thesis but I remained paralysed by indecision and grew paradoxically more lethargic and more restless by the day. My life felt intractably complicated but an expiring postdoc offer pressed for a decision. While I loved science and research, I was tired and battered from PhD life and was unsure about a research career. While I was eager for change of scene and pace, I was in a promising new relationship and wasn't ready to embrace a change.

A friend decided to read my tarot over Friday night beers, sure that she could help me decide once and for all if I was ready to move to Europe and start a research career. It turns out I wasn't, and I was relieved to find out.

Of course I don't attribute the fate of the cards with my decision, but rather I note that the act of participating in the reading was useful. My friend's earnest attempts to provide an accurate reading of the tarot cards prompted me to undertake a focused exercise is self-reflection, in which I acknowledged what I already knew. Through tarot, my thinking became visible and a decision was made with finality. In this sense, pseudo-knowledge can be useful, though clearly not for the reasons touted.

However, if we step back and look at scientific problems, or great societal challenges, like climate change, population health or energy production infrastructure, popularly understood pseudo-science approaches are truly revealed as lacking when compared to scientific claims. For example, the false knowledge claims (Lloyd 2014) central to the 2014 stoush over the validity of Australia's temperature records were limited in their usefulness to society. Many scientists invested in refuting the claims of fraudulent record keeping and an expert advisory group was impanelled to investigate these claims methodically, but in doing so largely repeated previous reviews of the methods used to derive the temperature data.

Hence, these sceptic knowledge claims were not enduringly useful to society. The subsequent reviews did not effectively prevent further sceptic criticisms, and arguably inhibited responses in Australia to a period of unprecedented heat. Conversely, the knowledge provided by the BoM's weather stations enables a complex response by society. These records reveal, far beyond our individual perceptions of the weather, Australia's experience of climate change and variability, and so permit us to respond meaningfully to the challenge of living in a highly variable and changing climate. In this sense, by adopting scientific approaches to understanding the physical mechanisms and impacts of climate change, we have posed a 'problem' that is useful and provides us with a vast capacity to as a society.

In accepting a spectrum of knowledge, scientists do not need to abandon falsifiability entirely, which provides a valuable element in assessing the utility of our spectrum of claims. This is demonstrated in applying a utility-centred approach to divergent claims about wind power. Claims that wind farms adversely affect human health through low-frequency sound that causes wind turbine syndrome do not need to be rejected outright as wrong, or denigrated as pseudo-science. Such claims can be useful if understood in the context of concerns about health, livelihoods and reluctance to embrace economic and social change. However, these criticisms of wind power are limited to providing insight into fear and worry.

By applying falsifiability, we see that wind turbine syndrome studies are typically undertaken by private companies with vested interests, have poor methodological controls, small sample sizes and are largely unfalsifiable (Hoepner and Grant 2015). Alternatively, claims derived from methodologically sound scientific studies of various energy sources are more broadly useful to society. For example, these studies provide us with information about the health risks associated with coal, wood fires and climate change. Robust studies can be used by society to make empowering decisions about balancing our need for stationary energy generation against the health risks of climate change, and hence are more useful because they enable a more enduring response.

The idea of viewing knowledge claims on a spectrum of legitimacy has been described as dangerous. In a *New York Times* Opinion piece (2013), philosophers Professor Massimo Pigliucci and Dr Maarten Boudry insist that there *must* be a clear demarcation between science and pseudo-science. While a small amount of pseudo-science is innocuous enough, a lot can be dangerous. Pseudo-science must not be allowed proliferate because of the significant implications for health and policy decision-making when a theory adopts the external facade of science, but fundamentally lacks scientific substance. Pigliucci and Boudry acknowledge that, of course, in practicality there can be no such sharp divide. Despite the inherent fuzziness of describing the world, they indicate that scientists must actively defend the gateways of wisdom, of systematic facts and rigorous thought from 'superstition and irrationality.'

I agree that the uncritical acceptance of the claims for pseudo-science can be dangerous to society; the uncritical acceptance of knowledge claims can lead to the well-meaning rejection of vaccines, or the spurious rejection of innovative technologies like wind farms by those with vested interests, or the wasteful rejection of useful and robust data, like that provided by the embattled BoM. Nonetheless, uncritically insisting on a flawed binary of knowledge claims cannot protect society from such dangers.

Since at least 1834, with the adoption of the collective name 'scientist' (Anon 1834), scientists have attempted to differentiate ourselves from non-scientists. In doing so, we have exalted scientific knowledge claims and rejected those of pseudo-science by using falsifiability as the core process of differentiation. However, as science has matured and expanded, scientists have overlooked incongruities in various contemporary branches of science. This process of differentiation from other knowledge claims uncomfortably excises climate science and various other contemporary or theoretically based disciplines from science. These are not be strictly categorised as science, but should not dismissed as pseudo-science.

In general, as a community, scientists have skilfully avoided confronting epistemological questions about how we acquire knowledge. Literary theorist Dr Peter Gaffney (2010) argues that 'it is not so much that science lacks access to questions concerning its own process, but that it actively renounces the conditions under which it would come to confront this process as an ontological question' (p. 10). I have attempted to confront these questions, and in response I argue that the definition of science, based on falsifiability, is limited. Rather than insisting on a dichotomy between these two forms of knowledge, scientists would be better placed to counter dubious knowledge claims if we view knowledge acquisition as a spectrum and evaluate claims based on how useful they are. This requires scientists to have confidence in the legitimacy and value of scientific knowledge, without it being necessarily underpinned by the alluring but limiting ideal of falsifiability.

Returning to my unpacking of the specific scientific nature of climate science and climate models, I propose that both the tools and knowledge of climate science are useful. Climate models are useful because of the change they *themselves* introduce into the world. Within the climate science discipline, climate models push us introspectively to the limits of our understanding of the physical climate system. Extrospectively, climate models are powerful, flexible tools that are useful to society because they provide us with the capacity to pose problems and, ultimately, to act.

NOTE

1. This approach was also described in *Chapter 2*.

GLOSSARY

Approximated   The process of estimating the true value.
Continuous equations   Mathematical equations where small changes in the input numbers results in small changes in the output values. These

are based on a continuous number line, unlike discrete mathematics where distinct, rather than continuous, numbers are central.

Discretisation The process of solving mathematical equations by using smaller steps and computations as a method of approximation.

Ensemble The use of slightly different models of the climate system together in order to understand aspects of uncertainty.

Emergent properties A property evident in a collection of components or complex system but which the individual members or parts do not have.

Empirical/empirically Based on observation and experiment, rather than theory or postulation.

External forcings Any influence on the climate that originates from outside the climate system itself. For example, changes in solar radiation and greenhouse gas concentrations are forcings.

Experiment The use of a climate model with a specific set of inputs and conditions. For example, an experiment using a climate model might aim to simulate the climate of the twentieth century, using a suite of changes in atmospheric composition, vegetation and solar radiation. See Taylor et al. (2012) for further details.

Falsifiability The knowledge claims of science are assessed through examining their falsifiability. *Any* contradictory instance to a scientific theory is sufficient to falsify that theory, regardless of how many positive examples appear to support it.

Fundamental natural laws Universal physical laws that are based on repeated scientific experiment and observations. Many fundamental physical laws are mathematical consequences of various symmetries of space, time, or other aspects of nature.

Internal variability Fluctuations in the weather and climate that are due to processes internal to the climate system, such as atmospheric and oceanic dynamics. The well-known El Niño-Southern Oscillation system is an example of internal climate variability.

Legitimacy The legitimacy of scientific knowledge is assessed through its falsifiability and adherence to key scientific concepts, such as the scientific method. This approach distinguished true science from the claims of pseudo-science, which are conversely, lacking in legitimacy.

Observations/observed Aspects of the physical climate system that are measured through instruments, such as thermometers, anemometers, rainfall gauges or satellites, rather than simulated using climate models.

Parameterisations The process of deciding on and defining the key aspects necessary to fully specific a model.

Problems Here I do not refer to problems in an ordinary sense. Rather, I refer to problems in terms of their capacity for usefulness as part of a creative process in research that allows us a deeper and richer connection to science.

Pseudo-science Here I use a precise meaning of pseudo-science, beyond the popular categorisation of beliefs such as astrology and homeopathy, which are mistaken for science, but are actually pseudo-science. Here, pseudo-science is more broadly a collection of practices that do not follow the scientific method or cannot be falsified, and hence lack true scientific status.

Simulation A single outcome of running a climate model experiment. A simulation (or realisation) of climate represents one possible pathway that the climate system might follow.

Skilful A term applied to climate models and simulations, referring to their ability to capture accurately key observed aspects of the climate system.

REFERENCES

Anon (1834) On the connexion of the physical sciences. The Quarterly Review 51:54–59.

Brown M (2014) Climate change deniers raise the heat on the Bureau of Meteorology. In: Sydney Morning Herald. http://www.smh.com.au/com ment/climate-change-deniers-raise-the-heat-on-the-bureau-of-meteorology-20140909-10eedk.html. Accessed 9 Apr 2015.

England MH, McGregor S, Spence P, et al (2014) Recent intensification of wind-driven circulation in the Pacific and the ongoing warming hiatus. Nature Climate Change 4:222–227. doi: 10.1038/nclimate2106.

Gaffney P (2010) The force of the virtual: Deleuze, science, and philosophy. University of Minnesota Press, Minneapolis.

Guillemot H (2010) Connections between simulations and observation in climate computer modeling. Scientist's practices and 'bottom-up epistemology' lessons. Studies in History and Philosophy of Modern Physics 41:242–252. doi: 10.1016/j.shpsb.2010.07.003.

Hagelaars J, Arndt DS (2013) My view on climate change. In: My view on climate change. https://ourchangingclimate.wordpress.com/2014/02/17/is-cli mate-science-falsifiable/. Accessed 6 May 2015.

Hargreaves JC, Annan JD (2014) Can we trust climate models?. WIREs Clim Change. doi: 10.1002/wcc.288.

Haughton N, Abramowitz G, Pitman A, Phipps SJ (2014) On the generation of climate model ensembles. Climate Dynamics. doi: 10.1007/s00382-014-2054-3.

Hawkins E, Edwards T, McNeall D (2014) Pause for thought. Nature Climate Change 4:154–156. doi: 10.1038/nclimate2150.

Hoepner J, Grant WJ (2015) Wind turbine studies: How to sort the good, the bad, and the ugly. In: The Conversation. https://theconversation.com/wind-turbine-stu dies-how-to-sort-the-good-the-bad-and-the-ugly-36548. Accessed 2 Nov 2016.

Huber M, Knutti R (2014) Natural variability, radiative forcing and climate response in the recent hiatus reconciled. Nature Geoscience. doi: 10.1038/ngeo2228.

IPCC (2007) Climate change 2007: The physical science basis. Contribution of Working Group I to the Fourth Assessment Report of the Intergovernmental Panel on Climate Change [Solomon S, Qin D, Manning M, Chen Z, Marquis M, Averyt KB, Tignor M and Miller HL (eds.)]. Cambridge University Press, Cambridge and New York, NY. 996 pp.

IPCC (2013) Climate change 2013: The physical science basis. Contribution of Working Group I to the Fifth Assessment Report of the Intergovernmental Panel on Climate Change [Stocker TF, Qin D, Plattner G-K, Tignor M, Allen SK, Boschung J, Nauels A, Xia Y, Bex V and Midgley PM (eds.)]. Cambridge University Press, Cambridge and New York, NY. 1535 pp. doi: 10.1017/CBO9781107415324.

Knutson TR, Tuleya RE (2004) Impact of CO2-induced warming on simulated hurricane intensity and precipitation: Sensitivity to the choice of climate model and convective parameterization. Journal of Climate. doi: 10.1175/1520-0442(2004)017%3C3477: IOCWOS%3E2.0.CO%3B2.

Knutson TR, Tuleya RE (2005) Reply. Journal of Climate 18:5183–5187. doi: 10.1175/JCLI3593.1.

Kosaka Y, Xie S-P (2013) Recent global-warming hiatus tied to equatorial Pacific surface cooling. Nature 501:403–407. doi: 10.1038/nature12534.

Lewis SC, Karoly DJ (2014) Assessment of forced responses of the Australian Community Climate and Earth System Simulator (ACCESS) 1.3 in CMIP5 historical detection and attribution experiments. Australian Meteorological and Oceanographic Journal 64:87–101.

Lloyd G (2014) Bureau of Meteorology 'adding mistakes' with data modelling. In: The Australian. http://www.theaustralian.com.au/news/nation/bureau-of-meteorology-adding-mistakes-with-data-modelling/story-e6frg6nf-1227048187480. Accessed 9 Apr 2015.

Lopez MA, Hartmann, DL, Blossey PN, et al (2009) A Test of the Simulation of Tropical Convective Cloudiness by a Cloud-Resolving Model. Journal of Climate 22:2834–2849. doi: 10.1175/2008JCLI2272.1.

Mueller P (2010) Constructing climate knowledge with computer models. WIREs Clim Change 1:565–580. doi: 10.1002/wcc.60.

Oreskes N, Conway E (2010) Merchants of Doubt, How A Handful of Scientists Obscured the Truth In Issues from Tabacco to Global Warming. Bloomsbury Press, New York.

Parker WS (2006) Understanding pluralism in climate modeling. Foundations of Science 11:349–368. doi: 10.1007/s10699-005-3196-x.

Parker WS (2011) When climate models agree: The significance of robust model predictions*. Philosophy of Science 78:579–600. doi: 10.1086/661566.

Parker WS (2013) Ensemble modeling, uncertainty and robust predictions. WIREs Clim Change. doi: 10.1002/wcc.220/full.

Petersen AC (2000) Philosophy of climate science. Bulletin of the American Meteorological Society 81:265–271. doi: 10.1175/1520-0477(2000) 081<0265: POCS>2.3.CO;2.

Pigliucci M, Boudry M (2013) The dangers of pseudoscience. The New York Times 21:1735–1741. doi: 10.1007/s00520-013-1720-z.

Popper K (1963) Conjectures and refutations: The growth of scientific knowledge. Routledge & K. Paul, London.

Roberts CD, Palmer MD, McNeall D, Collins M (2015) Quantifying the likelihood of a continued hiatus in global warming. Nature Climate Change 5:337–342. doi: 10.1038/nclimate2531.

Sanderson BM, Knutti R (2012) On the interpretation of constrained climate model ensembles. Geophysical Research Letters. doi: 10.1029/2012GL052665.

Schmidt GA (2007) The physics of climate modeling. Physics Today 60:72–73. doi: 10.1063/1.2709569.

Stott PA, Stone DA, Allen MR (2004) Human contribution to the European heatwave of 2003. Nature 432:610–614. doi: 10.1038/nature03089.

Stott PA, Allen M, Christidis N, et al (2012) Attribution of weather and climate-related extreme events. In: WCRP Position Paper on ACE. http://library.wmo.int/pmb_ged/wcrp_2011-stott.pdf. Accessed 16 Mar 2012.

Taylor KE, Stouffer RJ, Meehl GA (2012) An overview of CMIP5 and the experiment design. Bulletin of the American Meteorological Society 93:485. doi: 10.1175/BAMS-D-11-00094.1.

Tollefson J (2016) Global warming 'hiatus' debate flares up again. Nature. doi: 10.1038/nature.2016.19414.

Trewin B (2012) A daily homogenized temperature data set for Australia. International Journal of Climatology. doi: 10.1002/joc.3530.

UNFCCC (2016) Paris Agreement. FCCC/CP/2015/L.9/Rev.1, 1–27. United Nations Office, Geneva.

Weart S (2010) The development of general circulation models of climate. Studies in History and Philosophy of Modern Physics 41:208–217. doi: 10.1016/j.shpsb.2010.06.002.

CHAPTER 4

A Tribe of Scientists

Abstract Lewis examines the concept of credibility in scientific processes using examples of practising science during her PhD research. The scientific method provides scientists with a formalised approach to the generation of credible scientific knowledge. However, by drawing on personal experiences of scientific research, Lewis reveals a chasm between the subjective practice of science—so-called messy methods—and the sanitised version of methods in published works. Lewis explores contemporary social theory to described science as a nuanced, complex and messy process. Lewis proposes that such so-called mess, must be contained in published scientific works, including through the reporting of null results and the explicit description of scientific methods employed.

Keywords Palaeoclimate · Credibility · Messy methods · Null result · File drawer problem

The next entry in the assemblage of key scientific concepts is **credibility.** The **scientific method**, first discussed in *Chapter 1* as a formalised approach to the generation of credible scientific knowledge, is further explored through reflections on my own experiences of scientific research. I begin with a frank account of undertaking my PhD research. This was a frustrating experience that revealed a chasm between subjective scientific practices and objective scientific methods published in academic journals.

© The Author(s) 2017
S.C. Lewis, *A Changing Climate for Science,*
DOI 10.1007/978-3-319-54265-2_4

My own experience of practicing science during my PhD research was personal and intimate, although this experience was not reflected in any subsequent published accounts of my research.

By drawing on contemporary social theory, I describe science as a messy process that becomes effectively **sanitised** in published research outputs. This chapter continues the thread of the previous chapter, problematising differentiation as a means of affirming scientific legitimacy and credibility. I argue that science, like all research, is nuanced and complex in practice. Scientific practice would benefit from openly celebrating the plurality of approaches and experiences rather than cleansing them from published works.

SCIENCE IN PRACTICE

As an undergraduate science student, I was taught that scientific knowledge is acquired through the scientific method. Like physician Edward Jenner, the pioneer of immunisation discussed in *Chapter 1*, scientists use the scientific method to obtain scientific knowledge. This is the unifying process of observation, hypothesis, testing and theory by which scientific knowledge is codified. The scientific method also ensures that the observer—the scientist—is irrelevant. Following Jenner's method of developing a smallpox vaccination (Jenner 1798), *anyone* should be able to test its efficacy. Jenner's theory has been accepted not because of his authority as a great physician and scientist, but rather because of the durability of his results.

This approach to science sounds simple, fruitful and rewarding. Unsurprisingly, this isn't always the straightforward way in which science works. The complexities of scientific practice can be explored through my experience of enacting the scientific method during my PhD research. I did my PhD under 'non-traditional' supervisory arrangements. This is a familiar euphemism used in academic circles to convey that arrangements did not work out well between my supervisor and me during the early stages of my candidature. After an irreconcilable breakdown in our working relationship, I was left to carve out my own research path.

One positive outcome of this otherwise difficult situation was that I wrote my PhD thesis by the 'with publication' method[1]. I did not have a supervisor to guide me expertly through writing the traditional, long-form thesis or to provide feedback on my written work. By necessity, my thesis instead took the form of three articles published in peer reviewed journals. This route to a PhD had many benefits—I learned the process of academic publishing and, most importantly, had a small set of peer reviewed papers ready to add to my CV before I graduated.

However, this arrangement raised some complications. When the time came to put the three articles together into a thesis, a respected faculty member advised me that I needed some sort of thread that linked them together. Regardless of whether a student undertakes a traditional thesis or a thesis with publication, it should represent a coherent body of work and more than just a jumble of disparate fragments printed off and bound together. 'Easy,' I thought, 'that's me!' *I* was that linking piece, the thread that united all the elements of my research!

My PhD was messy. My 'freelance' status meant my PhD research was a circuitous adventure. I fell into an existing project that was tied to a major grant from an external funding body; I pursued my contribution to the grant outcomes eagerly, but thoughtlessly. I was researching by brute force, hoping that if I worked hard enough, something interesting would fall out of my enormous piles of raw data. Eventually something interesting did emerge, and after many months of research, I had obtained an enviable dataset. Next, I produced some fine looking plots of my scientific data that would become the seed of a peer reviewed journal article.

I was delighted to publish my first peer reviewed paper and I was excited that my first research foray raised more questions than it answered. These early, brute force scientific results were tantalising, but they were also equivocal. This first paper presented a record of climate variability in Indonesia around 30,000 years ago (Lewis et al. 2011). Since the early 2000s, the tropical regions have garnered attention from climate scientists as a particular area of interest. Around that time, an international group of researchers published evidence that the tropics are a far more dynamic player in the global climate system than previously thought (Wang 2001). We often associate the tropics with sunshine and warm humid weather, so it may come as a surprise that the tropical regions were ever involved in the ice ages.

The ice ages (**glacial periods**) are characterised not just by large ice sheets and prolonged cold conditions, but also by abrupt changes, including in the tropics. For many decades, **palaeoclimatologists** focused their research efforts on the slow waxing and waning of ice ages, until a paradigm shift in the late 1980s opened up research into these abrupt climate events. German scientist Dr Hartmut Heinrich (1988) discovered unusual layering in sediment cores taken from the Atlantic Ocean. These unusual layers were predominantly comprised of drop stones, yielded from the base of melting icebergs. Subsequently, the same debris layers were discovered in numerous sediment sites stretching across the Atlantic basin. Heinrich and other researchers meticulously reconstructed the trajectories of these ancient iceberg armadas, and eventually pieced together that during the cold glacial periods, huge icebergs periodically drifted across the North Atlantic. The discovery prompted many questions. What were huge icebergs doing travelling across the Atlantic? What was the climate doing during these times? What triggered this diaspora of ice?

Although scientists now have a name for these episodes of ice rafting—the **Heinrich events**—the *cause* of iceberg calving off the large northern ice sheets remains unknown. Their *impact* is, however, better understood. Scientists now know that these are semi-periodic events that recurred during the ice ages on average every 7,000 years, and that the Heinrich events had a very significant, abrupt impact on the climate system. The armada of icebergs that drifted into the North Atlantic eventually melted and flooded this highly sensitive ocean region with freshwater, which in turn perturbed global oceanic circulation patterns and led to very rapid cooling around Europe. The last Heinrich event plunged Europe into an abrupt chill around 16,000 years ago.

While scientists have now understood for some time that the high latitudes undergo very rapid and large climatic changes during the Heinrich events, the tropics were until recently considered comparatively dormant. The early 2000s research suddenly put the tropical regions into the spotlight. Dr Yongjing Wang (2001) and colleagues reconstructed changes in climate from chemical signals preserved deep in Chinese caves and demonstrated that the East Asian monsoon system was dynamic and highly changeable. The tropics are now considered an engine driving the global climate system, with the monsoons shifting vast amounts of heat and moisture up to the high latitudes. With huge numbers of people living within monsoon climates, such as in China, India and Indonesia, understanding the nature of monsoon changes has become an important research avenue (Partin et al. 2007; Cruz et al. 2009).

My first published paper was, at the time, a somewhat unusual contribution to the field of tropical palaeoclimatology (Lewis et al. 2011). It presented evidence from the understudied Indonesian region and provided very high-resolution and novel information about rainfall changes in the past. In the intervening years between the ground-breaking Chinese cave research and when my enthusiastic hands were let loose on a collection of Indonesian stalagmites (see Fig. 4.1), records of past climatic change in the tropics were being interpreted collectively rather than as

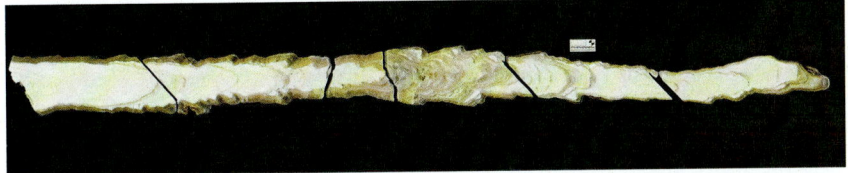

Fig. 4.1 Stalagmite (coded LR07-E1) from Luang Luar Cave in Indonesia provides evidence of climate change during Heinrich even 3.

single pieces of evidence. It was thought this set of evidence showed that changes occurring across the hemispheres were anti-phased. That is, during abrupt climate change occurring in response to the Heinrich events, the Northern Hemisphere monsoon region was much drier and the Southern Hemisphere monsoon affected areas simultaneously became wetter.

My research was centred on a particular Heinrich event—event 3, occurring roughly 30,000 years ago—and problematised this simplification. Together with the expertise of a collection of highly specialised scientists working under a federally funded grant, I showed that the spatial and temporal patterns of climatic changes during Heinrich event might actually be far more convoluted. These results indicated that sometimes areas in the Southern Hemisphere tropics around Indonesia will behave in the same way as Northern Hemisphere sites, and at other times they will be anti-phased. Each Heinrich event is likely to be unique.

A CIRCUIT BREAKER

After all that time slogging through Indonesian caves collecting samples and years in the laboratory, I was not much closer to understanding the climatic processes that were pertinent to my study area. My results hinted at a great complexity in climate responses to large-scale abrupt events. I was eager to discover more, so I leapt enthusiastically straight into the next study.

I wasn't yet sure precisely what I wanted to investigate, but I was aware that my first real foray into science had raised many more questions than it had answered. Was I interpreting my data correctly? Were these results only relevant to my study site on the Indonesian island of Flores or could they be applied more broadly to other regions? Were other researchers making valid assumptions by extrapolating climate information from

South America all the way across the Southern Hemisphere to Australia and Indonesia? I ended up employing vastly different approaches to test some of the assumptions I made in my first study (Lewis et al. 2010). Around this time, tensions in my candidature were bubbling away and my project began languishing. I became stressed, anxious and was seriously contemplating dropping out.

A circuit breaker came in the form of a small grant awarded by my university. I travelled to the NASA Goddard Institute of Space Studies (NASA GISS) to spend several months using their state-of-the-art climate model. I was there to model changes in the water cycle occurring during these Heinrich events (see Schmidt et al. 2014 for model details). It was exhilarating. I spent three months tinkering away, a FORTRAN language book in one hand and the other tapping away at my keyboard. My research was alive again; I was thrilled with the challenge and being pushed to think deeper and broader. Unable to sleep in the hot, humid craze of a New York City summer, I would stay up to the wee hours and debug code and make plots and write lists of ideas. I was so energised that I invited myself to return the following summer for another research stint.

My re-ignited enthusiasm paid off and I was eventually able to show that in some tropical regions, prevailing interpretations of past climatic changes were not necessarily the most accurate (Lewis et al. 2010).[2] In some locations, archives of climate change, such as stalagmites, are probably recording local climate signals, not regional or hemispheric-scale changes. In other locations, such as China, stalagmites are more likely to reveal large-scale changes. And in other locations again, like in West Africa, our new research suggested that the interpretation of chemical signals as showing long-term wetting and drying might be misguided. Instead, these locations could be picking up complex changes in the hydrological cycle. For example, rather than recording changes in rainfall at particular locations, natural archives from West Africa might be recording changes in where rainfall comes from, together with changes in circulation patterns *and* rainfall amounts (Tierney et al. 2011).

I was so proud of this research. It was intimately tied up with reconnecting with my research and restoring my passion for science. Midway through my PhD, I travelled back to the USA and proudly presented my results at a conference. After my talk, I was approached by a researcher who expressed a sense of relief that his particular field sites in Antarctica provided comparatively simpler study areas than the tropics I had just discussed. My mind raced. I hadn't thought this necessarily followed from my conference presentation, which was focused exclusively on understanding tropical climate signals and changes in the water cycle through time.

In response to this chance meeting, I leapt into my third study. This time, my research was focused on interrogating this assumption that the high latitudes record simple, readily interpretable climate signals (Lewis et al. 2013). Scientifically, this was the most challenging set of results to pick apart and understand. My co-authors and I could hint that complexity in the hydrological cycle was being preserved in ice core records of past climate change. That said, it was difficult to determine with confidence whether uncertainties in our results were from complexity in the climate system, or because of limitations in our experimental design. Even after I had submitted my thesis many months later, I wished that we could have conducted a more expansive set of climate model simulations than my collaborators and I had previously attempted. Despite these frustrations of unanswered questions, I was relieved—I had *finally* finished my PhD.

Methodological Hygiene

A thesis by publication requires introductory and explanatory material that explains how the different papers form a coherent body of work, in a similar fashion to an exegesis for creative or artistic PhDs. Looking at my nearly finished thesis, it seemed obvious that I had defined and united my research. I was the element that turned three disparate papers into a meaningful 'body of work.' Nonetheless, such an explanation would not have sufficed; science follows only the scientific method. Science is science regardless of the scientist, and the irrelevancy of any individual researcher is widely considered a great strength of science and its potential for discovery. It was not possible to weave through my individual papers a thread that was defined by me—the researcher—and present this as a PhD thesis for formal examination. This flexibility does not exist under the conventions of a contemporary science PhD programme.

Instead, I dutifully complied with our far less narcissistic, traditional view of research. I retrospectively fitted order to my research methods. Hypotheses were posited, objectives were met. Each piece became a carefully constructed study that fed seamlessly into the next. My PhD looked as though it was meticulously planned from the beginning, with no hiccups, false starts or frustrated tears. There was no student-supervisor breakdown, which was awful and traumatic but also counter-intuitively fostered a feisty and productive exchange of contentious and exciting ideas. There was no formative trip to New York City to learn a whole new approach and a desperately needed reconnection with my discipline. There was no chance meeting at a conference that set the path for a new and ambitious line of inquiry.

There was, in short, no life in this document. It was as though all the sparks of process had been hidden by impersonal language. I recoiled. I hated the finished product (and I still hesitate to drag my thesis off my bookshelf, cringing when asked to do so). Rather than creating something *more* meaningful and useful from my research experience, this sanitised version seemed to me a far *less* valuable account of my research.

The practice of removing all skerricks of humanity from research outputs is standard. Science ultimately seeks to make unqualified statements about reality. To do so, scientists both passively and actively reject the elements that don't quite fit in. We exclude all the invisible work that creates our research but does not seem worth recounting. We reject the details of our research that we assume everyone already knows; various other aspects of research, such as the personal, inconclusive or unremarkable, are repressed for one reason or another.

Social scientist John Law (2004a) describes this as a process of **Othering**,[3] whereby we separate out these things that don't quite fit from those elements that neatly fit in and are retained. By this process of Othering, all the qualifying statements are concealed, and processes, assumptions, instruments and skills are standardised. It is the 'subjective' and the 'personal' that tend to disappear first. In writing my PhD thesis and retrospectively fitting order to my experiences, I came to feel as though the largest part of my work—both the subjective and personal—had been deleted.

Such exclusions and deletions from scientific methods are expected, but are they always inherently useful? Law argues that traditional research methods provide a system for offering bankable guarantees on invested research time. These orthodox research approaches guide us quickly to our destination—methods allow us to learn that particular hypotheses are correct, or that particular approaches are flawed. Methods act as 'a set of short-circuits that link us in the best possible way with reality' (Law 2004a, p. 10). Traditional research methods are useful and important for conducting research. Streamlined methods, as exemplified by the scientific method, can and do lead to solid results, reliable papers and productive careers.

However, Law (2004b) suggests that in embracing *only* these approaches, we are practicing a form of methodological hygiene:

> Do your methods properly. Eat your epistemological greens. Wash your hands after mixing with the real world. Then you will lead the good research life. Your data will be clean. Your findings warrantable. The product you will produce will be pure. Guaranteed to have a long shelf-life. (p. 510)

Law argues that, for the social sciences at least, this culture of research practice can have dangerous implications. In adopting conventional, risk-averse research approaches, we have given ourselves over to mechanical replacement. Our research effectively becomes automated, limiting intellectual inquiry. In response, Law (2004a) provides a broader definition of **methods**. In this definition, research methods are not simply a set of techniques or even a philosophy of method, and:

> It is not even simply about the kinds of realities that we want to recognise or the kinds of world we might hope to make. It is also, and most fundamentally, about a way of being. (p. 10)

Methods become fundamentally important as researchers *are* methods; methods are our way of working and our way of being beyond the laboratory and field sites.

The same limitations of methodological convention that Law explicates for social inquiry can equally be explored in the physical sciences. Do the traditional academic methods of inquiry—our idealised scientific method —capture 'the mess, the confusions, the relative disorder' of what we actually study? My own account of my PhD research demonstrates that scientists cannot be neatly separated from the messy, confused and disordered research. I embraced standard scientific practice and clean, 'hygienic' approaches by methodically and systematically removing myself from my own PhD.

These deletions matter. First, concealing our messy research methods is not merely a matter of the politics of research itself, it is a matter of the politics of reality. While the scientific method ideally represents a short-circuit to the generation of credible scientific knowledge, the denial of mess during its implementation can abrade credibility. In sanitising our methods, in giving ourselves over to mechanical replacement, we essentially deny elements of reality. As Law (2004a) states,

> If the world is complex and messy, then at least some of the time we're going to have to give up on simplicities. . . . if we want to think about the messes of reality at all then we're going to have to teach ourselves to think, to practice, to relate, and to know in new ways. We will need to teach ourselves to know some of the realities of the world using methods unusual to or unknown in social science. (p. 2)

In producing my PhD thesis, I was not simply denying aspects of my research, but by consequence I was denying aspects of reality.

Confronting Our Mess

These acts of sanitising have implications for scientific knowledge beyond elusive allusions to the nature of reality. Research *needs* to be **messy** simply because that's the way research actually is. Scientific accounts of methods must collectively reflect this, at least to some degree. Secreting these 'ragged ways in which knowledge is produced in research' (Law 2004a, p. 19) in official accounts limits scientific inquiry in tangible ways. In the physical sciences, this systematic Othering—these quiet absences from accounts without acknowledgement—is demonstrated in our response to the **null result**. Researchers are notoriously reluctant to share negative results. Researchers and academic journals alike tend to cast their results as a story that we believe others will want to read. Negatives findings are left unpublished and equivocal findings inflated, with our words carefully selected to emphasise the importance of a result. As a result, we eventually read of the 'alarming' rather than 'modest' increase in obesity.

Sanitising (or Othering) in sciences becomes subtly evident in the resulting peer reviewed publications through this cultural interpretation of results. A recent article in *Science* magazine explored one particular aspects of the reluctance of scientists to share null results and their surreptitious conflation of results (Couzin-Frankel 2013). Couzin-Frankel (2013) notes that medical researchers offer diverse explanations for why they failed to report data on an outcome they had previously pledged to examine; one researcher considered that their results were 'just uninteresting and we thought it confusing so we left it out,' and another that when 'I take a look at the data I see what best advances the story, and if you include too much data the reader doesn't get the actual important message' (p. 69). As a result, these 'uninteresting' medical science results went unpublished or unreported, though the interviewed researchers typically believed such practices of exclusion improved their published work.

However, these inflections reveal a bias in our official accounts of science, which is commonly referred to as the 'file drawer' problem. I do not use bias to indicate prejudice to one specific scientific viewpoint over another. Rather the bias rests in the fact that science creates accounts of research—such as my PhD thesis—that values only some parts of our methods and work practices. For example, the selective reporting of statistically significant results and withholding of insignificant results (those analytical results that scientists selectively leave in the file drawer) is a 'pernicious form of publication bias' (Franco et al. 2014, p. 1502). Franco et al. (2014) argue that this biased reporting of results is problematic for two key reasons. First, researchers waste time, effort and resources in conducting studies that have already been done, but have not been publicly reported. Second, if researchers conduct similar studies to those previously executed but not reported, and obtain significant results by chance, then the published literature erroneously reflects only the chanced upon significant result.

Biased reporting inhibits our collective ability to assess the state of knowledge in a field or on a particular topic because null results are simply not published. In political scientist Franco et al.'s (2014) study focused on biased reporting of significant results, 15 of 25 participants reported that they had abandoned projects because they believed that null results have no publication potential even if they themselves found the results interesting. If making such knowledge public is not an outcome of our research methods, how can conventional scientific methods adequately serve their intended purpose of generating credible scientific knowledge?

The file drawer problem is just one of a set of cultural practices that shape the production of scientific knowledge. This set of practices has been illuminated by sociologists Bruno Latour and Steve Woolgar, who studied scientists at length for their book *Laboratory Life: The Construction of Scientific Facts* (1979). This seminal ethnography of science details their two-year observation of the scientists of the Salk Institute in San Diego and describes a 'tribe of scientists.' Law's (2004a) interpretations can help us understand this concept of a **tribe**:

> Scientists have a culture. They have beliefs. They have practices. They work, they gossip, and they worry about the future. And, somehow or other, out of their work, their practices and their beliefs, they produce knowledge, scientific knowledge, accounts of reality. (p. 19)

Science *is* a set of practices, and these practices are intimately shaped by our historical, organisational and social context. This is a **constructionist** approach. Scientific knowledge is *constructed* within those practices. Scientific knowledge does not evolve in a vacuum; scientists participate in the social world, 'being shaped *by* it, and simultaneously *shaping* it' (Law 2004a, p. 12). This is not to say that scientific knowledge is simply invented by scientists, but rather acknowledges that the process of building scientific knowledge is constructed in scientific practice, which in turn is situated within broader social, political and economic circumstances.[4]

As part of an adherence to cultural practices, scientists tend to delete the largest part of scientific work. As scientists, we delete that which we consider unimportant and preserve only a small subset of our scientific practice in our methods sections. There is always a complexity and mess that is pared away in favour of polished and neat research outputs, including in climate science research.

In *Chapter 3*, for example, I discussed the messiness of climate models as a source of scientific knowledge. Climate models represent that physical climate system, but they are a tangle of complex physics and intricate approximations of human decisions. Climate models incorporate many sources of uncertainty and integrate at least some degree of arbitrariness. Of course, when results from climate models are presented in scientific journal papers, they reflect our collective set of practices and present a subset of this information in a conventional, 'scientific' style. Law's 'taken-for-granted assumptions, instruments, or skills' (p. 36) have been concealed and information that is not considered important goes unnoticed.

This is not to say that information is nefariously hidden from scrutiny, or that scientists dishonestly practice science by concealing aspects of scientific processes and methods. Scientists do not actively select results that fit a forgone conclusion and delete others. Rather, essential or useful elements of practice are lost through sanitising methods, and practices stifle creative lines of inquiry. As Law (2004a) notes, 'Conventional talk of 'method' is closely associated with rules and norms for best practice. Indeed, though method is usually more than this, it sometimes becomes indistinguishable from lists of do's and don'ts' (p. 40).

On occasion, the messiness of research methods is made publicly explicit. The Overly Honest Methods (#overlyhonestmethods) phenomenon was a popular trend on social media. This light-hearted initiative collates honest descriptions of methodologies that would never find a place in our peer reviewed published literature. In these descriptions, the basis for

Jesse McDonald @McDoogled   ⚙ **Follow**

"Sample size was smaller than planned because I had been in grad school for 10 yrs & my advisor wanted me to graduate." #overlyhonestmethods

Ben Seymour @benosaka   ⚙ **Follow**

Blood samples were spun at 1500rpm because the centrifuge made a scary noise at higher speeds. #overlyhonestmethods

Georgina Guedes @georginaguedes   ⚙ **Follow**

"We assume 50 Ivy League kids represent the general population b/c actual 'real people' can be sketchy or expensive." #overlvhonestmethods

Clint Mercuur @MercuryMan91   ⚙ **Follow**

The samples incubated at ambient temperature in a remote border customs office for 5 months #overlvhonestmethods

Caitlyn Cardetti @CaitlynCardetti   ⚙ **Follow**

Experimental time points were chosen so I didn't have to come into lab in the middle of the night or over the weekend. #overlyhonestmethods

Sylvain Deville @DevilleSy   ⚙ **Follow**

we didn't read half of the papers we cite because they are behind a paywall #overlyhonestmethods #OA

Fig. 4.2   Examples of Overly Honest Methods (#overlyhonestmethods) shared on Twitter.

analytical decision-making is bluntly provided. Figure 4.2 shows examples from a cursory Google search, which are fairly typical of displays of methodological honesty.

From my own PhD research I could add:

- One sample was selected for further analysis because I couldn't find the other, far more interesting, sample until months later when I eventually found it on a dusty windowsill in the sample preparation room. I had placed it there while juggling boxes of sample vials, marker pens and lumps of rocks.
- We obtained 10 times as many measurements as necessary because we thought the sample was really old and valuable. It turns out it wasn't.
- Weight measurements were obtained for only one in every two samples. I tried to weigh every sample, but a poorly timed sneeze sent several carefully weighed out grains of stalagmite material flying off the scales and into the ether around me.

I love these honest methods, both funny and familiar. In addition humour and familiarity, this inventory of honesty elucidates Law's list of 'do's and don'ts.' Experimental designs or analytical approaches determined by happenstance can yield productive and useful outcomes, but there are few arenas where this is acknowledged as part of scientific research. As a tribe, scientists more or less practice restrictive and risk-averse methodological hygiene.

Remaking Methods

Law implores social researchers to unmake their sanitised methodological habits. Rather than craving certainty and expecting that research can arrive at more or less stable conclusions about how the world is, social scientists should broaden, subvert and remake method. Researchers ought to divest themselves of their distracting concern with these hygienic, sanitised approaches and instead embrace the multiple, diverse and the uncertain.

Such entreaties to broaden and loosen methodological approaches would be rather more perplexing for the physical sciences. For physical scientists, the acknowledgement of 'mess' in methods could easily put a scientist in a bind: if I do try to capture the mess, my research may appear 'messy.' And surely 'messy' is really just a euphemism to describe being poorly done? For a physical, rather than social, scientist discussion of remaking methods likely seems unnecessary or uncomfortable. When a friend reviewed an early version of this chapter, the margins were littered with 'this all sounds a bit narcissistic,' or 'NO results in the methods section!' or 'methods MUST always produce the same results.' These well-meaning and frank comments typify a culture in which a scientist must employ specific approaches in order to have their work taken seriously and be funded.

How then can we remake the scientific tribe's culture, beliefs and practices to allow methods that acknowledge and accept the reality of research? Beyond Law's epistemological treatise, there remains the issue of how scientists can start to embrace mess in a practical sense. In order to reject the simplistic association between 'mess' and poorly done, we first must whole-heartedly embrace messiness.

This begins with methodological transparency, and Franco et al. (2014) proposed solution to the 'file drawer' problem. They recommend a two-stage review system for studies. The first stage of review presents the experimental design and pre-analysis plans for consideration and the second presents the results. That is, studies must be pre-registered for results to be subsequently published and should be complemented by incentives not to hide away insignificant results in file drawers. Publication of pre-analysis plans and registries themselves will increase researcher access to null results. The second stage of review is simply our traditional publication model. Under this approach, a group will register a proposed study and analytical methods, for example, to investigate the impact of a particular diet in mice on a specific disease outcome. The group must then publish their final results, regardless of the significance of the outcomes, as a statistically insignificant result (a so-called weak result) may prove as valuable and informative to the broader discourse as a 'strong' result that provides specific insight into the targeted disease.

However, this two-stage process still requires a strictly organised way of researching that does not readily accommodate flexible research methods. Some argue that science also requires a shift in cultural norm, whereby scientists embrace and share results without regret or hesitation. Daniele Fanelli studies bias and misconduct at the University of Montreal and insists on methodological honesty. In Couzin-Frankel's (2013) exploration of negative results, Fanelli argues 'the only way out of this [is that] people report their studies saying exactly what they did' (p. 69).

How would this work in practice? As a community, scientists accept that methodological choices can be legitimately justified by the rather antiquated sounding approach of 'personal communication' (or 'pers. comm.'). Using the phrase 'personal communication' in a published article, a researcher can make a shorthand argument that a particular decision was appropriate because of a personally communicated (rather than peer reviewed) authority (e.g. Lewis et al. 2013). In this way, a conversation with another researcher can legitimate an aspect of a methodological approach. If such conversations are relevant to scientific approaches, I ask the provocative question—why then can't we 'pers. comm.' ourselves?

In *Chapter 6*, I will detail my experience with a paper I submitted to a journal. The paper endured an unusually arduous peer review processes, which provides insight into this hypothetical question. After my paper had been through three exhaustive rounds of peer review, I noticed that over these successive rounds of review, one particular reviewer had demanded that I provide a reference for nearly every sentence in the methods section of the study. I could appreciate comments requesting that my introduction and discussion be more tightly grounded to the ongoing discussion occurring in the wider scientific community. But in the methods section?!?! How could I possibly explore new areas or demonstrate new ideas if I was bound so tightly by approaches already used? In this example, I was not coveting 'overly honest methods,' but 'just the right amount of honest methods.'

To fulfil Fanelli's suggestion that scientists report what they actually did, in my paper:

- I used this particular experimental set-up because I thought it would be interesting. It was.
- I did not conduct a second set of experiments because I could not afford the time or money to do so. The results were still interesting and provide a good basis for scientific discussion.
- I used a much earlier version of the climate model than the current release and I will continue to do so because I've been writing this paper for 2 years on Sunday afternoons. On Monday to Friday, I have to work on my paid job.

Many of the concepts I have discussed throughout this chapter may at first seem elusive or irrelevant when removed from Law's social science context and applied to the work of scientists. However, in practice, an openness to mess does not necessarily require a distinct set of research practices, but rather a change in the way that scientists relate to and talk about their research. After all, I inadvertently explored mess during my PhD research, an experience that was certainly not highly automated or risk-averse. However, at the same that I researched long-term tropical climate change, I also learnt how to participate in the scientific tribe and apply its cultural norms, which resulted ultimately in my sanitised thesis. While I enacted messy methods during my PhD research, I simply never had space in which to talk formally about this exploration of mess.

In collectively considering the messiness of scientific research and methods, I don't expect that we will soon see climate science PhD theses in which short auto-ethnographies detailing the experience of research connect a series of peer reviewed scientific papers. There is, however, room between the orthodox scientific method and auto-ethnographic accounts, which science collectively would gain from acknowledging.

By celebrating uncertainty, diversity and indefiniteness within research, we could save a lot of time and effort by reducing unnecessary repeated experimental endeavours. In addition, an openness to mess has the potential to expand research directions and limits. Science, like all research, is nuanced. In practice, it does not typically follow a singular method, a simplified progression, nor a simple process of observation, hypothesis, observation and theory. While the complex fabric of reality and our connection to it need not become the dominant focus of published papers, this must be reflected *somewhere* in scientific work. While those of us who practice or support science may find comfort in the uniformity of established methods and approaches, at some point adopting only this approach inhibits our ability innovate. If relying on one singular, streamlined process of knowledge production is limiting, celebrating messy approach may encourage high risk but potentially high reward research, and hence will not diminish the generation of credible scientific knowledge, but rather enhance it.

NOTES

1. Other universities describe this mode as 'by publication,' 'by papers' and 'by compilation.'
2. As the publication process for my first PhD paper took a significant amount of time, the publication date (2011) was pushed later than this second study (2010), which I began later.
3. Throughout this chapter, I draw on Law's ideas to analyse the scientific method and research process, though his is the work of a social scientist.
4. This broader context is discussed in further detail in *Chapters 5* and *6*.

GLOSSARY

Constructionist The concept that humans generate knowledge and meaning from an interaction between their experiences and their ideas.

Credibility The extent to which science is recognised as a reliable and trustworthy source of information, particularly in terms of how the research adheres to traditional scientific principles, such as the scientific method.

Glacial periods An interval of time characterised by substantially lower global temperatures and glacial advances. These cold periods (i.e. the ice ages) were interrupted by shorter periods that were substantially warmer, called the interglacials.

Heinrich events A natural phenomenon occurring during the ice ages when large icebergs break from Northern Hemisphere glaciers and traverse the North Atlantic. The melt water from icebergs acts to disrupt oceanic and atmospheric circulation and causes large-scale climatic change.

Methods From Law (2004a), methods is not simply a technique for producing research output, 'it is not just a philosophy of method, a methodology. It is not even simply about the kinds of realities that we want to recognise or the kinds of world we might hope to make. It is also, and most fundamentally, about a way of being' (p. 10).

Messy   This is an imprecise description that encompasses the vague, diffuse, ephemeral and elusive nature of reality and undertaking research.

Null result   A scientific result that does not support a hypothesis.

Othering   The process by which researchers exclude information that does not fit into an approach as irrelevant.

Palaeoclimatology/palaeoclimatologists   The study of past climate change and variability/those studying past climate change and variability.

Sanitised   This is the process through which research analysis and methods are transformed from a messy, real-world activity and presented as a streamlined procedure.

Scientific method   An approach of systematic and repeated observation, measurement, experiment and the formulation, testing and modification of scientific hypotheses.

Tribe   Latour and Woolgar (1979) description of scientists as bound together by a set of practices. This is summarised by Law (2004a) as 'Scientists have a culture. They have beliefs. They have practices. They work, they gossip, and they worry about the future. And, somehow or other, out of their work, their practices and their beliefs, they produce knowledge, scientific knowledge, accounts of reality' (p. 19).

References

Couzin-Frankel J (2013) The power of negative thinking. Science 342:68–69.

Cruz FW, Vuille M, Burns SJ, et al (2009) Orbitally driven east–west antiphasing of South American precipitation. Nature Geoscience 2:210–214. doi: 10.1038/ngeo444.

Franco A, Malhotra N, Simonovits G (2014) Publication bias in the social sciences. Unlocking the file drawer. Science 345:1502–1505. doi: 10.1126/science.1255484.

Heinrich H (1988) Origin and consequences of cyclic ice rafting in the Northeast Atlantic-Ocean during the past 130,000 years. Quaternary Research 29:142–152.

Jenner E (1798) An Inquiry Into the Causes and Effects of the Variolae Vaccinae: A Disease Discovered in Some of the Western Counties of England, Particularly Gloucestershire, and Known by the Name of the Cow Pox. Printed for the author, by Daniel Nathan Shury, London. 1801.

Latour B, Woolgar S (1979) Laboratory Life. The Social Construction of Scientific Facts. Sage Publications, Beverly Hills.

Law J (2004a) After Method: Mess in Social Science Research. Routledge, London.

Law J (2004b) Making a Mess with Method. Centre for Science Studies, Lancaster University, Lancaster LA1 4YN, UK, http://www.lancaster.ac.uk/fass/resources/sociology-online-papers/papers/law-making-a-mess-with-method.pdf.

Lewis SC, LeGrande AN, Kelley M, Schmidt GA (2010) Water vapour source impacts on oxygen isotope variability in tropical precipitation during Heinrich events. Climate of the Past 6:325–343. doi: 10.5194/cp-6-325-2010.

Lewis SC, Gagan MK, Ayliffe LK, et al (2011) High-resolution stalagmite reconstructions of Australian–Indonesian monsoon rainfall variability during Heinrich stadial 3 and Greenland interstadial 4. Earth and Planetary Science Letters 303:133–142.

Lewis SC, LeGrande AN, Kelley M, Schmidt GA (2013) Modeling insights into deuterium excess as an indicator of water vapor source conditions. Journal of Geophysical Research: Atmospheres. doi: 10.1029/2012JD017804.

Partin JW, Cobb KM, Adkins JF, et al (2007) Millennial-scale trends in west Pacific warm pool hydrology since the Last Glacial Maximum. Nature 449:452–455. doi: 10.1038/nature06164.

Schmidt GA, Annan JD, Bartlein, PJ, et al (2014) Using palaeo-climate comparisons to constrain future projections in CMIP5. Climate of the Past 10:221–250. doi: 10.5194/cp-10-221-2014.

Tierney JE, Lewis SC, Cook BI, et al (2011) Model, proxy and isotopic perspectives on the East African Humid Period. Earth and Planetary Science Letters 307:103–112. doi: 10.1016/j.epsl.2011.04.038.

Wang YJ (2001) A high-resolution absolute-dated late pleistocene monsoon record from Hulu Cave, China. Science 294:2345–2348. doi: 10.1126/science.1064618.

CHAPTER 5

The Nature Peepers

Abstract Lewis focuses on expertise and authority in science. This discussion explores several key questions about scientific process, including how science identifies important research questions, assesses the utility of research and evaluates scientific expertise. Using experiences of peer review and recent controversies in the scientific community, such as the hiatus in global temperatures, Lewis examines the evaluation of scientific knowledge and the place of commercial publishing in research. Lewis problematises the distinction between the scientific expert and non-expert, and recommends several practical and conceptual step towards making science and its processes more accessible. These include embracing open access publishing and two-stage peer review, and recognising knowledge produced from citizen scientists.

Keywords Expertise · Authority · Citizen science · Hiatus · Open access publication

This chapter focuses on **expertise** and authority as key assemblage concepts. I examine key elements of **governance** in science, including how we identify important research questions, assess the utility of research and evaluate scientific expertise. I begin with a detailed discussion of the process of **peer review** using my own experiences of publishing scientific papers, which highlight issues around how knowledge gains credibility.

© The Author(s) 2017
S.C. Lewis, *A Changing Climate for Science*,
DOI 10.1007/978-3-319-54265-2_5

This exploration opens out into a broad discussion of the implications of commercial academic publishing, using key examples from recent debates within the climate science community. Finally, I explore the role of the **citizen scientist** as a contributor to scientific knowledge.

With a focus on the practitioners of science, I consider what gets said in science, who can access that information and who is allowed to contribute to science. Introducing the idea of a critical **contract** between science and society, I argue that this relationship must be renegotiated in response to the changing nature of expertise and authority in science. Rather than reinforcing barriers between expert and non-expert, science should seek to make scientific processes of enquiry and its ensuing results more accessible to the non-expert. Similarly, science must recognise the value of other types and sources of knowledge as credible, including the value of scientific knowledge derived from 'non-experts.'

An Ungoverned Murmuration

I've always loved the natural world. In earlier chapters, I described my childhood forays into astronomy, taxonomy, botany, zoology, geology and meteorology. I explored on foot and by bike, collecting animal and plant parts. I also explored with my microscope, binoculars or telescope, collecting interesting sightings. I tried to work out what things were and what things were for. Later, I undertook a long period of highly specialised and formalised scientific training as an undergraduate and PhD student. In the introductory chapter I referred to this training as an 'apprenticeship.' Gazing back on my career, what does this subsequent period of formal training mean for my childhood experiences of science? Does this formalised and dedicated training mean I was *not* practicing science during my childhood explorations of the world? Or does training mean that I am more of a scientist now? While at first glance such questions might seem purely personal reflections, these simple questions reveal an unclear distinction between scientists and non-scientists that requires further consideration.

Science is intuitively understood as a particular approach to acquiring knowledge, with a shared set of practices. While it is evident that science is *something*, it is hard to pin down and define science in a way that adequately encompasses all the pursuits and practices that we readily recognise as 'science' and exclude those activities we regard as 'non-science.' In *Chapter 3*, I problematised the separation of science and non-science knowledge claims through falsifiability and pseudo-science. Here, I extend this exploration to the *practitioners* of science by exploring the notion of expertise in science. Who has enough expertise to be in Latour and Woolgar's (1979) **tribe** of scientists? What precisely are the means by which scientists distinguish themselves from non-scientists? What precisely are the means through which science comes to consider particular knowledge claims expert?

Philosopher and scientist Professor Michael Polanyi (1962) attempts such a differentiation between expert and non-expert by characterising science as a **republic**. He suggests a 'community of scientists is organised in a way which resembles certain features of a body politic and works according to economic principles similar to those by which the production of material goods is regulated' (p. 54). Polanyi (1962) proposes that scientists constitute a community, by contrasting a group shelling peas with a group of scientists going about their research. The pea-shellers are working on the same task of shelling at the same time, but their individual efforts are not coordinated in a way that scientists are. Although scientists often work individually (like the pea-shellers), we are in fact coordinated to some degree because we make continual adjustments based on the results of all others (which the shellers do not).

Polanyi suggests that these continual adjustments made by scientists are mutual and hence that the pursuit of science ultimately occurs by independent self-coordinated initiatives, which is the most efficient possible organisation of scientific progress. He argues that any attempt to organise scientific practice under a single **authority** would be futile. Doing so would negate independent initiatives and reduce our joint effectiveness to that of just a single governing person or body. In Polanyi's view, we can think of scientists as a **murmuration** of starlings, a huge flock, wheeling and darting in a tight but fluid formation. Like starlings, scientists are coordinated, communicating rapidly with each other over long distances and networks, with a change in one group member capable of affecting all others animals within the group, but with no single leader determining the flock's movements.

Although I love the idea of an enormous flock of scientists in collective and dynamic flight, it is a naïve characterisation of contemporary science. This conceptualisation suggests both that scientists respond intuitively to all others within our field of influence, and that all scientists exert equal influence on our fellow starlings. Such interpretations of scientists are far too clearly defined—scientists are a flock of starlings, and the passing, stray cockatoo is perhaps a social scientist, quite clearly morphologically and immediately different and distinguishable. And so too is the wayward pigeon flapping by, which could be a scorned pseudo-scientist who is distinctly, definitively outside of science. However, as I discussed in depth in *Chapter 3*, distinguishing scientific knowledge claims from those that are not scientific is not straightforward. Furthermore, it is also evident that science must operate within some form of **governance** and complex power structures. In reality, not every starling has the same capacity to affect change as the next. The Honours, PhD and postdoc starlings, the starlings from developing countries or from less prestigious research institutions simply do not have the same field of influence.

As part of these power structures, science (among other academic pursuits) invests deeply in the process of exclusion of non-experts, erecting and fortifying barriers that exclude other forms of knowledge. Polanyi (1962) discusses how expert contributions to scientific knowledge are evaluated as such:

> The first criterion that a contribution to science must fulfill in order to be accepted is a sufficient degree of plausibility. Scientific publications are continuously beset by cranks, frauds and bunglers whose contributions must be rejected if journals are not to be swamped by them. This censorship will not only eliminate obvious absurdities but must often refuse publication merely because the conclusions of a paper appear to be unsound in the light of current scientific knowledge. (p. 57)

Polanyi's 'cranks, frauds and bunglers' must be excluded from the scientific flock—our murmuration—as it ducks away from 'obvious absurdities.' Such rigid exclusions prompt fundamental questions about who is excluded and who makes such decisions.

Professor of Climate and Culture Mike Hulme (2009) disputes claims of ungoverned science, arguing that there is a form of governance within science, whereby scientists identify important research questions, assess the utility of research and evaluate expertise. In the remainder of this chapter, I explore elements of this system of governance: peer review processes; publication models; and citizen science. That is, I consider what gets said in science, who can access that scientific information and who is allowed to contribute.

Becoming Science

Peer review is a critical process through which scientific knowledge gains authority (Hulme 2009). The process of peer review firmly underpins modern scientific practice and is used to exclude various knowledge claims from science. When scientific research has conducted been and novel results written into a manuscript, an author submits a paper to an academic journal for consideration of publication. An editor assesses the paper and if the quality and aims of the study align with those of the journal, it is next sent to a set of scientific peers for a thorough assessment of the quality of the research. This is ideally the unbiased exercise through which the merit of a scientific contribution is assessed by others within the scientific community based on accuracy, importance and the intrinsic interest of its subject matter (Polanyi 1962).

A journal article that has successfully survived peer review conveys assurance that accurate scientific knowledge of considerable importance is contained within. However, my own experiences of peer review do not, by any means, reflect an unbiased and rigorous process of improvement. I have experiences on multiple faces of the peer review process—I regularly act as a peer reviewer for a number of journals, work as an editor for an Australia climate journal, and I have a sufficient number of published papers that I am no longer acutely jealous when another early career researcher gets a paper accepted by a high-profile journal.

This suite of personal experiences provides a range of insights into peer review, with one particular drawn out example of peer review contradicting Polanyi's (1962) view that this is unbiased exercise of evaluation with the common goal of bettering scientists' collective knowledge. This particular review battle was a vicious, personal and self-interested fight that raged for well over 12 months. The skirmish started when I submitted a paper using multi-disciplinary approaches to a journal that would provide readership in both disciplinary communities. From the point of submission, at roughly three-month intervals, I received new rounds of comments that the reviewers had provided to the journal editor. When an email update would appear in my inbox overnight I periodically turned from a mostly normal and reasonable human, into a raging, twitching, muttering mess.

The first round of reviews provided mixed opinions. Two reviewers who shared a common disciplinary background liked the study and the approach used, though they were a little ambivalent about the conclusions and thought the writing was unnecessarily muddled and long-winded. I was pleased, until I read on. The third reviewer did not like my analysis and thought there was little value in my approach. In response, I substantially revised the manuscript and was pleased to produce a far clearer study as a result. I returned my revised manuscript to the journal editor and waited optimistically.

The second round of peer review provided even more polarised opinions. The first two reviewers (from the previous round) were satisfied with the changes I made after considering their comments, and indicated that the paper was ready for publication, after some minor tweaks. The final reviewer was a new participant in the battle, and categorically did not like my work. The reviewer described my work as 'hand-waving,' 'imprecise,' 'long-winded,' and 'dull.' The reviewer seemed to insist that I rewrite my study to exclude all of my ideas or approaches and cite only literature from their disciplinary area. I prevaricated. Finally, some weeks later, I capitulated. I sat down and I re-wrote my paper, ending up with a manuscript that would have made Tolstoy proud. It was a meandering epic of textbook proportions. There were highs, there were lows, there was war and there was peace. None of it made sense.

Disappointed at producing a manuscript devoid of purpose, I sat down and re-wrote my paper again. This time, I attempted to distinguish between reviewer comments that were useful and those that I believed to be purely self-motivated or stylistic. I submitted my paper again, and it went to review again. After revisions, I submitted my paper again, it went to review again and eventually the editor grudgingly accepted my paper for publication. Despite this final success of publishing the paper, I maintain that the clarity of the study was degraded through the process of peer review.

I am clearly not an impartial observer in probing the shortcomings of the reviews of my own manuscripts. However, such experiences are hardly unusual. In 2015, an indisputable demonstration of self-serving reviewing and sloppy editing came to light when a manuscript was submitted to the journal *PLOS ONE* focusing on the attrition of female participants in the biological sciences (see Bernstein 2015). A reviewer sparked controversy by suggesting to the two female authors that 'It would probably be beneficial to find one or two male biologists to work with (or at least obtain internal review from, but better yet as active co-authors), in order to serve as a possible check against interpretations that may sometimes be drifting too far away from empirical evidence into ideologically biased assumptions.' The reviewer went on to suggest that 'it might well be that on average men publish in better journals ... perhaps simply because men, perhaps, on average work more hours per week than women, due to marginally better health and stamina.'

The suggestions that the female researchers' work would be strengthened by the addition of one or two male co-authors sparked outrage.

Twitter responded with customary fury and inspired the hashtag #AddMaleAuthorGate (Woolston 2015), which has permeated academic vernacular. The journal responded by apologising for the review, removing the review from its database and submitting the paper to a fresh set of reviewers. Furthermore, the academic editor who handled the manuscript was asked to step down from the journal. Nonetheless, the vastly inappropriate review comments highlight systematic problems with peer review, particularly by revealing that much of the value of the review system is dependent on the implicit assumption of high-quality, responsive and responsible editors.

I should note that in the lottery of peer review, I have also experienced excellent, skilled people with excellent, skilled ideas reviewing my papers. These editors and reviewers have taken time out of their tight schedules to give consideration to my work. They have contributed thoughtful comments that have quite rightfully changed the way I have interpreted a dataset, or structured a manuscript, or they have provided invaluable insight into the appropriateness of certain statistical approach. In these cases, they have enriched my work and my capacity as a young scientist to research, and I have enjoyed the process of improving my own work with the help of these generous people.

However, both own underwhelming experiences of peer review and the very public #AddMaleAuthorGate example of intractable reviewers and a neglectful editor are familiar to most scientists. On the other hand, the obverse experience of attentive editors and reviewers who begin with the expectation that they will enjoy and value a paper is equally familiar to most scientists. This myriad of ways in which peer review is applied to manuscripts casts doubts on the usefulness of the peer review process. I have described, with intent, the process of peer review as combative. It *feels* like a battle. In my particular example, I felt that I lost a battle. I felt that the quality of work was diminished, as I spent considerable of time producing a longer, meandering and less precise manuscript. More importantly, I felt that science lost. The peer review process did not enhance the strength of the science nor the clarity with which it was communicated.

While a reliance on peer review process to underpin scientific publishing and knowledge acquisition is appealing, I argue that such distorted instances of review demonstrate peer reviews its limitations and that the idea of a democratic republic of science is ultimately false. The review process did not provide an unbiased assessment of the accuracy, systematic important and the intrinsic interest of my scientific contribution. Instead, it appears that the reviewer whose comments I vehemently disputed had a greater influence on the editor and final outcomes than any of the other reviewers, authors or associate editors.

Within our hypothetical murmuration of starlings, the capacity to influence is clearly not equitably distributed, and there are clearly sites of power within science; there is power within, and some sort of governance of our murmuration of starlings. While I have criticised elements of peer review, I have not yet made recommendations for improving the process nor distinguished the key differences between poor and quality reviewing. As peer review is interlinked with the application of power and influence of science more broadly, I will provide suggestions of responses to these quandaries later in this chapter.

Big Business

A broader facet of authority influencing the scientific **academy** can be seen in traditional publication models, and particularly the place of high-profile journals in the generation of scientific knowledge. Under a traditional model of academic publication, the majority[1] of scientific papers are published in subscription-based journals (Van Noorden 2013). I have already described parts of this model, with research conducted, written into a manuscript, submitted to a journal, reviewed and published. Academics perform peer review work on an unpaid basis—a hugely time-consuming voluntary contribution. In addition, a journal's editorial board is usually comprised of well-respected researchers, who typically work on a voluntary basis, although occasionally are paid a small sum for their contributions.

When the lengthy review process has been finished and an editorial decision has been made, the manuscript is ideally accepted for publication, and the author will be sent an invoice for publishing costs. These can be quite substantial, with charges for colour images and excess pages rapidly becoming costly and often running to several thousand dollars. A typeset and copyedited manuscript is typically published both online and in print format, although I do not know of anyone who still reads print editions of journals. At this point, the article is secreted away behind a pay wall and can only be accessed through payment.

University libraries often buy large, cost appealing packages of subscriptions to journals. These packages are a bowerbird's nest of multiple, and often unrelated, journals and back-editions that may or may not be useful to a university's researchers. As specifics of the costs of these packages are generally subject to strict nondisclosure agreements, the value of such bundles is debatable (Mayor 2004; Lemley and Li 2015). Furthermore, journal list prices have risen steadily, with recent estimates suggesting that commercial publishers, who hold an enormous marketshare in academic publishing, yield profit margins of roughly 20–30%[2] (Van Noorden 2013). These subscription packages are not universally affordable, with enormous disparities in journal access between universities and between countries, with many journal papers hidden well beyond the financial capacity of many researchers. Inequities in journal access can be particularly large in the Australian research sector, where universities rely heavily on government funding which is allocated differently to different universities.

To summarise this simplified view of the traditional publication model, the public fund the great majority of academic research through government research grants, through which scientists are paid their salaries and project costs to conduct research. Scientists then pay to publish journal articles through page charges, volunteer their time to the publishers as reviewers and finally, scientists pay to read each other's research through their institution's library subscription fees.

In addition to profiting from the traditional model, some journals arguably exert great influence on driving the direction and focus of scientific research and research practice. The most prestigious scientific journals—the British publication *Nature* and its sibling *Science* from across the Atlantic—are hugely influential. These are both high-impact publications and an acceptance letter from either is highly coveted, essentially making careers in a single act. The prestige of such journals is irresistible to scientists; I can't help but yearn for results that would be attractive to either publication. Scientists always aim to publish in the 'best' journals possible, and these high-impact publications are undoubtedly the 'best' in terms of career pay off, given the prescriptive metrics that are applied to evaluating success in an academic career. Under intense pressure to generate novel and exciting results, researchers who succeed in publishing high-profile papers revel in kudos and career success. The same cannot be said for the persistent researcher who tenaciously devotes years to a well-conceived, replicated experiment with solid evidence of a useful result. Popping champagne bottles and offers of prestigious tenured positions do not await the doggedly determined researchers.

At the same time as coveting a high-profile publication under my name, I can also see that the role of high-profile journals in science is murky for several reasons. In competitive and fast moving scientific fields, the desire to avoid being scooped by an opposing laboratory can come to outweigh the risk of issuing a retraction at a later point in time. In this way, the dominance and power of these prestige journals can also influence scientific practice. A pertinent example of mishap and misconduct in research was revealed by the website Retraction Watch (2015). A high-profile *Science* paper claiming that short, personal conversations could change people's minds on the issue of same-sex marriage was published in 2014 (LaCour and Green 2014), but retracted following revelations the data were faked by the junior co-author LaCour, who was on his way to a prestigious assistant professor position at Princeton, riding the wave of his high-profile, high-impact paper. While I do not attribute blame for LaCour's misconduct to *Science*, big rewards for high-profile papers and increasing pressure to publish frequently can act to create environments in which misconduct thrives.

Both *Science* and *Nature* have a similar capacity to set disciplinary agendas, deciding whether scientific research is not just worthy but also interesting. At first pass, this does not sound troubling—of course the most prestigious journals are interested in the most interesting, cutting edge and exciting work! But what happens when a single editorial group has the power to encourage large groups of researchers to lurch into new directions?

A recent example of this editorial power might be revealed in the recent discussion of the **hiatus**, an observed slowdown in the rate of global warming at the Earth's surface. Around the time that the IPCC handed down their fifth assessment report (IPCC 2013), there was heated public discussion about the hiatus. Some contributors to the public debate were adamant that this was unequivocal proof that the world was not warming at all and that global warming was unfounded. Prior to 2013, a small number of scientific papers identified the hiatus as an area of research interest and interrogated the physical causes behind the change in the rate of observed warming (Meehl et al. 2011). However, around 2013, the number of comments and papers centred on the hiatus ballooned. It seemed as though every month, a hiatus-focused paper from the *Nature* subsidiary *Nature Geoscience* was published and countered rapidly by *Nature Climate Change*, or vice versa (Kosaka and Xie 2013; England et al. 2014; Huber and Knutti 2014).

Throughout this dynamic period of 2013–2015, the state of the climate system was intensely debated. In responding to feisty public comments that the hiatus disproved global warming, many climate scientists countered that they never claimed that the rate of global warming would be linear, with the same increase in heat recorded in each successive decade. Rather, complexity should be *expected* within the climate system. In this avalanche of academic interest, the hiatus was categorically attributed in the published literature to a multitude of causes, including deep ocean processes, aerosols, measurement error and the cessation of ozone depletion. Editorial (2014, p. 157) reminded us that the 'average rate of warming at the Earth's surface is only one piece in the climate change puzzle' and hence does not reveal complexity in the system and is not singularly insightful. I also argued this point in a newspaper commentary piece published in the lead up to the release of IPCC's 2013 report (Lewis 2013). The point is so obvious, so simple, that we must ask—if the hiatus was so unsurprising, then why did it precipitate a blizzard of high-profile research papers? Who decided it was a new and exciting avenue of research that should be covered in depth?

The hiatus discussion was being influenced and directed from multiple locations, including sceptic blogs, the mainstream media and scientists, as well as high-profile journals. The widespread interest in the hiatus over these years has been explored previously, including in Lewandowsky et al.'s (2015) paper, which focused on the hiatus as an example of the inadvertent intrusion of memes that arose outside the academic community into scientific discourse and thinking. Such intrusions are termed 'seepage' in reference to how climate denial has influenced the scientific community. While sceptic influences are certainly powerful participants in climate discourse that should not be underestimated, I argue that the predominance of discussion within prestige publications, whether through editorials or expert comment or research articles, set the hiatus research agenda and started a bandwagon rolling.

My interest in the hiatus rivals a teenager's interest in an overplayed pop song—stretched to the point of irritation. For other climate scientists, it is fodder for a long-term research agenda. Perhaps the hiatus was a bandwagon worth rolling and by studying its precise characteristics, scientists will elucidate important aspects of the climate system. Regardless of personal investment and interest, the uncritical acceptance of scientific interest in the hiatus is problematic. While I love *Science* and *Nature,* and the routine of sitting down to a luxuriously long coffee break once a week and catching up on their letters and research articles, I am nonetheless uncomfortable with an essentially commercial enterprise[3] having the capacity to exert so much influence over scientific research agendas.

The processes of our current publication models clearly encompass a multitude of influences, powers and authorities that are not always transparent or acknowledged. While we posit that peer review is a reliable mechanism through which the value of a scientific contribution is made, the authors, reviewers and editors of submitted papers clearly work within the deep constraints of a largely commercial enterprise that is itself subject to public and political whim. While such processes seem deeply entrenched, and these contradictions between ideals and outcomes impossible to reconcile, there are several practical approaches for improving the outcomes of scientific review and publication that we can collectively adopt.

As a Retired Engineer

Hulme (2009) proposes that the long-standing relationship between science and society constitutes a **contract**, through which science provides society with one form of guidance for decision-making. In high-stakes research fields with great social and political interest, such as climate science, science's contract with society is more difficult to define and negotiate, and requires adjustments to society's differing expectations of scientific information. In order to meet the obligations of our contemporary contract with society, Hulme proposes we must democratise science and allow 'greater lay expertise to operate in the governance of science' (p. 81). That is, science cannot continue to rely solely on claiming greater expertise in making peer reviewed knowledge claims, an approach that is problematic for several reasons.

It is not a trivial task to work out who is in our flock of scientists, and who is excluded. Hulme also problematises a reliance on expertise simply because 'the expertise of scientists and the claims of scientific knowledge do not exhaust the sources of expertise or authority to which society may turn in seeking guidance for the decisions that must be made' (p. 80). In understanding and responding to climate change, society requires guidance from all variety of experts, including those in the social and political sciences, and economists, engineers and technicians.

I propose several conceptual and practical responses the challenge of modern scientific publishing and practice that I have discussed, and Hulme's challenge of democratising science and its governance. We can begin with a conceptual shift in thinking about the authority of scientific knowledge, which for me begins in my childhood experience of collecting and collating the Australian Alps, piece by piece, and my earlier questions about how such activities align with science. Were these not the experiences of a scientist because they were informal, fleeting, purposeless experiences, conducted outside the lab? A gentle solution can be found during that famed meeting described by Anon (1834), where the terminology of 'scientist' was adopted, but other alternatives, were raised:

> Others attempted to translate the term by which members of similar associations in Germany have described themselves, but it was not found easy to discover an English equivalent for natur-forscher. The process of examination which it implies might suggest such undignified compounds as nature-poker, or nature-peeper, for these naturae curiosi, but these were indignantly rejected.

The description **nature peeper** encapsulates a wonderful, antiquated and delightful image of being permitted to steal quick glances at nature, but never quite see its entire, elusive form. It also hints at the idea that participating in science is, at least to some degree, intuitive and accessible. While I present the term lightheartedly, such a category usefully describes the science of untrained scientists. As a child, I did not have sophisticated language or approaches or knowledge to frame my thoughts. However, I was arguably participating in the same beautiful and rewarding shared endeavour, with the same, and somehow mutual goals and values that I participate in today as a professional scientist.

Indeed, climate science increasingly recognises the practical value of nature peepers. Knowledge generated outside orthodox scientific expertise by citizen scientists (Gura 2013; *Nature* Editorial 2015) is being folded into and incorporated into peer reviewed papers. In these cases, enlisting non-expert volunteers allows researchers to investigate otherwise very difficult problems, for example, when the research would have been financially and logistically impossible without citizen participation. These are large networks of 'amateur experts' who help to collect and analyse scientific data in collaboration with a researcher.

Several of my scientist colleagues have been involved in projects that require the assistance of volunteers. The OzDocs project, involves volunteers digitising early records of Australian weather from weather journals, government gazettes, newspapers and our earliest observatories (Fenby and Gergis 2013). This project provided a better understanding of the post-1788 climate history of southeastern Australia. Personal computers also provide another great tool for citizen collaborators. In one ongoing project, climate scientists conduct experiments using publicly volunteered distributed computing (Pall et al. 2011). Participants agree to run experiments on their home or work computers and the results are fed back to the main server for analysis. Oxford University has now produced over 100 million years of climate model data using otherwise idle computer time. These data have allowed researchers to gain a better understanding of climate models and extreme climate events.

There are other practical steps forward for more broadly engaging society in scientific pursuits, including through changing scientists' approach to and expectations of peer review. Climate scientists often discredit knowledge claims from outside science as irrelevant because they have not undergone the necessary peer review process, but I have already shown how peer review does not always adequately assess the merit of knowledge claims. If peer review is the process by which authority is established, then it *must* be fit for purpose. I do not suggest that traditionally peer reviewed science is necessarily junk and should be treated as such, and nor do I suggest that we altogether abandon peer review. Rather I argue that keeping the peer review process and decisions hidden away behind barriers is limiting for both the practitioners of science and science's relationship with society.

I have no doubt that scientists can keep the most valued elements of peer review *and* make it more accessible. For example, several years ago I submitted a paper to an open access journal with a two-stage review process, one open to critique by the general public and one by my scientific peers (Lewis et al. 2010). This approach aims to use the full potential of the Internet to foster scientific discussion and enable the rapid publication of scientific papers. In my experience, this worked well. This simple change is a way to open up obscured academic processes to the outside world, to incorporate the knowledge of other experts, and to provide accountability within the review process. Experimentation with the creation of expert scientific knowledge will strengthen scientists' ability to determine a paper's quality and suitability for publication, including, for example, curbing the inappropriate demands of a rogue reviewer to include a male author would be less likely.

In another practical step towards democratisation of science, publication models are shifting monumentally in response to factors such as the 'death' of print media, the uptake of social media and drive for open access. Discussion of open access publication as an alternative to the traditional publication model has increased in recent years. Under this open system, the authors of a paper pay a higher cost for the publication of a paper, which is then made publicly available at no cost to the reader. This model is gaining traction. Some researchers have outright rejected commercial publication models, arguing that the exclusive hiding of taxpayer-funded research behind pay walls is 'immoral' (Van Noorden 2013). Others have rejected the traditional model on financial grounds.

In April 2012, Harvard University started encouraging its faculty members to make their research available freely through open access publication (Sample 2012). This coincided with a widespread call to boycott journals published by the Elsevier published house. If one of the richest universities cannot afford to support the pay wall model of publication, what chance is there for less fortunate researchers with access only to comparatively impoverished libraries? Commercial journal profiteering has incited strong responses. In 2011, Alexandra Elbakyan, a Russian researcher, launched Sci-Hub, a site that made 48 million journal articles freely available online (Bohannon 2016). In doing so, and refusing to withdraw Sci-Hub, she has been at the receiving end of protracted lawsuits, including from publishing monolith Elsevier. Elbakyan's open access dance is ongoing, characterised by technological move and countermove to maintain access to the articles for researchers in institutions and countries where subscriptions are simply unaffordable.

There has been a steady increase in the uptake of open access publication, particularly as various funding bodies demand that research papers based on publicly funded research should be free for all to read. A suggested benefit is that your research is more likely to be widely read and cited, than if squirrelled away behind a prohibitively expensive subscription. I note that open access publication remains polarising. Subscription-based journals are pushing back on open models or offering mixed mode publication options for the same publication. These publications tend to argue that the costs of publication are worth it, because they maintain high-quality research and oversee the publication process. Others argue that open access is an intergenerational sore point, in which more established researchers have the career security and financial capacity to make ethical decisions around publication, a luxury which may be unavailable to younger researchers.

While open access remains a live wire in academic tearoom discussions, there are many intermediate options for academics to respond to feelings of unease, or incense, at peer review itself. First, scientists can respond by being better reviewers by providing detailed reasons for suggestions and critiques, being precise about what its expected and what is are merely suggesting, and being clear with the editor about our motivations as a reviewer. Scientists also have a degree of autonomy in choosing which journals to submit papers to. I appreciate that while not all are accessible financially, there are many different types of journals, including those operated by scientific societies that are typically maintained and operated by committed scientific members and publish for free for the benefit of the disciplinary community.

In addition, authors can demand accountability in the review process. In 2015, I submitted a paper to a journal that I had held in high regard. My paper was returned with a one-line rejection from a reviewer and a flippant editorial note that likened my work to 'beating a dead horse.' I lodged a complaint with the publishers, have refused all subsequent requests to review for this journal and 'blacklisted' it from my catalogue of appropriate venues for my work, in favour of those journals that provide attentive editors and considerate, critical reviewers.

Finally, the academy can more willingly acknowledge that scientists are increasingly expected to be active contributors to society. Currently, most universities apply very narrow criteria for evaluating staff performance, focusing primarily on the number and quality of peer reviewed publications. In this way, individuals are reduced to a small set of numbers called h-indices and impact factors. An expanded evaluation of performance and achievement that included impact on and involvement with society would encourage scientists to engage directly with, rather than exclude, non-experts.[4]

Professor Stuart Firestein (2012, p 14) asks, 'How does anyone even get started being a scientist? And if it's intimidating to trained and experienced scientists, what could it be to the average citizen? . . . . Is this the reason that science can seem so inaccessible?' It is not just the necessary technical aspects and specific language of scientific research that makes it inaccessible to the untrained non-expert but also the separation of the expert from the non-expert by science. These delineations could be blurred by rewarding academics for their time engaging with the public. If public engagement were viewed a *fundamental* part of academic work, as a core responsibility, rather than a fringe activity, science would more broadly accessible. I have spoken at schools and libraries and National Science Week activities, which are all good fun, with small groups of people excited to talk to a climate scientist about my views on the many facets of climate change. These commitments occur on the periphery of my 'real' work according to standard university norms, but this view only acts to separate scientists from everyone else and reinforce exclusivity in scientific knowledge production and dissemination.

At first the idea of relinquishing elements of peer review, scientific publication and expertise sounds daunting. Would embracing guidance and involvement from a variety of experts and non-experts be chaotic? Would science be beset by Polanyi's (1962) 'cranks, frauds and bunglers' and their 'absurdities'? At first it might seems so. For many climate scientist, an appearance in the media prompts unsolicited advice and comments from the general public. Typically after I give a radio or television interview, I receive a flurry of emails or handwritten letters about my work. While a few genuinely desire further information, most begin with 'As a retired engineer....' and contain page after page of notes and equations that purportedly disprove my scientific research on climate change.

These letters calling on 'retired engineer' status as a form of expertise are commonly received by climate scientists, and commonly irritating. I usually dismiss them as nonsense—Malcolm the retired engineer from Warrandyte in Victoria[5] cannot be an expert because *I* am the expert. While such repudiations are tempting and feel entirely warranted, they are not necessarily the most useful conceptual understanding of differences of opinion about scientific information. More specifically, if we focus again on the concept of usefulness explored in *Chapter 3*, I can argue that my scientific results are useful to society, having been peer reviewed, scientifically accepted and based on a body of preceding scientific work. In contrast, Malcolm's hypothetical manifesto in my inbox detailing that the world is actually cooling is based on dubious data appearing on a blog, and analysed using equations and methods that have not been explained or published. Hence, his results are rejected, but not *only* because of the expertise conferred through *my* scientific training.

My proposed practical and conceptual adjustments to scientific processes—open access publication, citizen science projects and multi-stage peer review—are already occurring, and have not beset scientific research with Polanyi's cranks. Instead, such changes in scientific processes have made scientific processes and results more accessible to non-experts. These changes strengthen the relationship of science with society, and more broadly liberate climate science from unattainable demands of being an exhaustive source of knowledge to society.

NOTES

1. The estimates of the percentage of publications that are open access vary considerably, and depend on the definition of open access. Van Noorden (2013) used Scopus citation information to determine that in 2011, 12% of articles were fully open access. An alternative approach used by Björk et al. (2010) put the number at 24%. A further data mining approach by Archambault et al. (2013) suggested that 48% of the literature published in 2008 was available for free in December 2012.
2. In 2015, reports indicated that academic publisher Elsevier earned about $1.58 billion in profit on about $9.36 billion in revenue (Peters 2016).
3. While Science is published by the nonprofit American Association for the Advancement of Science (AAAS), it is a commercial enterprise, with pay-walled articles. Meanwhile, the Nature Publishing Group (NPG) is privately owned and does not disclose its financial results.
4. In the following chapter, I will expand on this idea of engagement and explore the differing approaches to climate scientists in their interactions with policymaking and politics.
5. This is not based on any real letter or person.

GLOSSARY

Academy The academy is the institution and community concerned with the pursuit of research and scholarship.

Authority Here I refer to science as an authority, having societal influence through our knowledge claims.

Citizen scientist These are volunteers who work in collaboration with scientists to expand scientific data collection and analysis.

Contract A concept of Hulme's (2009) to describe the critical relationship between science and society, which is bound by a contract of understanding and obligation.

Expertise Credibility and knowledge in a particular area obtained by study, training or formal experience.

Governance This refers to the way in which is the way the rules, norms and actions are structured and imposed within a group or organisation, including informally through outside influences.

Hiatus The slowdown in the rate of global surface warming over the period of approximately 1997–2015.

Murmuration A flock of starlings is called a murmuration, a mass of birds that appear to be connected together in flight.

Nature peepers   An alternative term to 'scientist' described by the *Quarterly Review* (1834). Here, I use this term to describe science that has not been formalised and is, at least to some degree, intuitive and accessible.

Peer review   The disinterested process by which the merit of a scientific contribution is assessed within the community based on accuracy, systematic important and intrinsic interest of its subject matter.

Republic   This is Polanyi's (1962) description of science, comprised of a 'community of scientists [is] organised in a way which resembles certain features of a body politic and works according to economic principles similar to those by which the production of material goods is regulated' (p. 54).

Tribe   Latour and Woolgar's (1979) description of scientists as bound together by a set of practices. This is summarised by Law (2004) as 'Scientists have a culture. They have beliefs. They have practices. They work, they gossip, and they worry about the future. And, somehow or other, out of their work, their practices and their beliefs, they produce knowledge, scientific knowledge, accounts of reality' (p. 19).

References

Anon (1834) On the connexion of the physical sciences. The Quarterly Review 51:54–59.

Archambault E, Amyot D, Deschamps P, et al (2013) Proportion of Open Access Peer-Reviewed Papers at the European and World Levels 2004–2011. Science-Metrix Inc.

Bernstein R (2015) PLOS ONE ousts reviewer, editor after sexist peer-review storm. In: Science. http://news.sciencemag.org/scientific-community/2015/04/sexist-peer-review-elicits-furious-twitter-response. Accessed 13 Jun 2015.

Björk B-C., Welling P, Laakso M, et al (2010) Open access to the scientific journal literature: Situation 2009. PLoS ONE 5:e11273–9. doi: 10.1371/journal.pone.0011273.

Bohannon J (2016) The frustrated science student behind Sci-Hub. Science 352:511–511. doi: 10.1126/science.352.6285.511.

Editorial (2014) Hiatus in context. Nature Geoscience 7:157–157. doi: 10.1038/ngeo2116.

Editorial (2015) Rise of the citizen scientist. Nature 265. doi: doi:10.1038/524265a.

England, MH, McGregor S, Spence P, et al (2014) Recent intensification of wind-driven circulation in the Pacific and the ongoing warming hiatus. Nature Climate Change 4:222–227. doi: 10.1038/nclimate2106.

Fenby C, Gergis J (2013) Rainfall variations in south-eastern Australia part 1: Consolidating evidence from pre-instrumental documentary sources, 1788–1860. International Journal of Climatology 33:2956–2972. doi: 10.1002/joc.3640.

Firestein S (2012) Ignorance: How it Drives Science. Oxford University Press, New York.

Gura T (2013) Citizen science: Amateur experts. Nature 496:259–261.

Huber M, Knutti R (2014) Natural variability, radiative forcing and climate response in the recent hiatus reconciled. Nature Geoscience. doi: 10.1038/ngeo2228.

Hulme M (2009) Why We Disagree About Climate Change: Understanding Controversy, Inaction and Opportunity. Cambridge University Press, New York.

IPCC (2013) Climate change 2013: The physical science basis. Contribution of Working Group I to the Fifth Assessment Report of the Intergovernmental Panel on Climate Change [Stocker TF, Qin D, Plattner G-K, Tignor M, Allen SK, Boschung J, Nauels A, Xia Y, Bex V and Midgley PM (eds.)]. Cambridge University Press, Cambridge and New York, NY. 1535 pp. doi: 10.1017/CBO9781107415324.

Kosaka Y, Xie S.-P (2013) Recent global-warming hiatus tied to equatorial Pacific surface cooling. Nature 501:403–407. doi: doi:10.1038/nature12534.

LaCour MJ, Green D P (2014) When contact changes minds: An experiment on transmission of support for gay equality. Science. doi: 10.1126/science.1259329.

Latour B, Woolgar S (1979) Laboratory Life. The Social Construction of Scientific Facts. Sage Publications, Beverly Hills.

Law J (2004) After Method: Mess in Social Science Research. Routledge, London.

Lemley T, Li J (2015) "Big deal" journal subscription packages: Are they worth the cost?. Journal of Electronic Resources in Medical Libraries. doi: 10.1080/15424065.2015.1001959.

Lewandowsky S, Oreskes N, Risbey JS, et al (2015) Seepage: Climate change denial and its effect on the scientific community. Global Environmental Change 33:1–13. doi: 10.1016/j.gloenvcha.2015.02.013.

Lewis S (2013) Debunking the persistent myth that global warming stopped in 1998. In: The Age. http://www.smh.com.au/environment/climate-change/debunking-the-persistent-myth-that-global-warming-stopped-in-1998-20130927-2ui8j.html. Accessed 9 Apr 2015.

Lewis SC, LeGrande AN, Kelley M, Schmidt GA (2010) Water vapour source impacts on oxygen isotope variability in tropical precipitation during Heinrich events. Climate of the Past 6:325–343. doi: 10.5194/cp-6-325-2010.

Mayor S (2004) US universities review subscriptions to journal "package deals" as costs rise. British Medical Journal 328:68–0. doi: 10.1136/bmj.328.7431.68.

Meehl GA, Arblaster J M, Fasullo J T, et al (2011) Model-based evidence of deep-ocean heat uptake during surface-temperature hiatus periods. Nature Climate Change 1:360–364. doi: 10.1038/nclimate1229.

Pall P, Aina T, Stone DA, et al (2011) Anthropogenic greenhouse gas contribution to flood risk in England and Wales in autumn 2000. Nature 470:382–385. doi: 10.1038/nature09762.

Peters J (2016) "Everyone" Downloads Research Papers Illegally. In: Slate. http://www.slate.com/articles/health_and_science/science/2016/04/science_magazine_can_t_defend_its_flawed_business_model.html.   Accessed 13 Sep 2016.

Polanyi M (1962) The republic of science, its political and economic theory. Minerva 1:54–74.

Retraction Watch (2015) Author retracts study of changing minds on same-sex marriage after colleague admits data were faked. In: Retraction Watch. http://retractionwatch.com/2015/05/20/author-retracts-study-of-changing-minds-on-same-sex-marriage-after-colleague-admits-data-were-faked/. Accessed 20 May 2015.

Sample, I (2012) Harvard University says it can't afford journal publishers' prices. The Guardian. https://www.theguardian.com/science/2012/apr/24/harvard-university-journal-publishers-prices. Accessed 16 Jun 2015.

Van Noorden R (2013) The true cost of science publishing. Nature 495:426–429.

Woolston C (2015) Sexist review causes Twitter storm. Nature 521:9. doi: 10.1038/521009f.

CHAPTER 6

Into the Hinterland

Abstract Lewis presents new descriptors for scientific practice. By juxtaposing two disparate views of what it is to be a scientist and what it means to apply scientific methods, Lewis questions objectivity as core tenet of science. Lewis contrasts her own reflexive science with a colleague's deeply held belief that objectivity is essential. Next, Lewis provides a broader discussion of recent climate science controversies, and the role of scientists in policy and decision-making. Lewis argues that the expectations of science are changing and a greater transparency of practice is required in response. Lewis proposes the idea of the hinterland, a new conceptual terrain in which scientists can acknowledge the culture, beliefs and practice of science with greater transparency.

Keywords Objectivity · Transparency · Post-normal science · Blogging · Conceptual hinterland

This chapter proposes new key scientific concepts. Before I present a new set of scientific concepts, I unpack an existing scientific assemblage entry—**objectivity**. This chapter questions objectivity as core tenet of science by juxtaposing two disparate views of what it is to be a scientist and what it means to apply scientific methods. I contrast my own experience of a reflexive and intimate connection to science with a colleague's deeply held belief that objectivity is essential to science. This small tension

© The Author(s) 2017
S.C. Lewis, *A Changing Climate for Science*,
DOI 10.1007/978-3-319-54265-2_6

between **subjective** and objective leaning scientific colleagues opens into a broader discussion of the role of objectivity in scientific practice. Should scientists advocate? Should scientists participate actively in the production of policy and decision-making processes, or would this compromise an essential impartiality?

Given the limitations of the more familiar concepts of knowability, legitimacy, credibility and authority in scientific practice I have discussed throughout, in this chapter I suggest a new set of concepts that scientists can incorporate into their work. I call these concepts a **hinterland**, which is a new conceptual terrain that scientists can explore. Within this terrain, scientists can acknowledge the culture, beliefs and practice of science that influence our ways of working. This new focus in scientific practice begins with the idea of **transparency.**

A Diversity of Practices

For the four years of my PhD candidature, I shared a revolving door office with a nebulous group of students. We were all undertaking vastly different environmental science projects, using very different techniques and approaches, and we had very different ideas about our PhD research. We also had vastly different understandings of science. One officemate, who was (and remains) my dearest friend, loved facts. She was an excellent analytical scientist, driven by an intense desire for order and a need to understand how each piece of data slotted together. For her, science was the pursuit of facts and scientists were the collectors who turned data into fact. To her, science is, by definition, objective. Inevitably, we would end up in heated arguments about science and objectivity.

She would insist that scientific data speak for themselves and scientists must allow them to do so. She believed that science is best conducted when dispassionate, to the extent she has argued that PhD projects have the best outcomes when a student does not even much like their research topic. For her, to *enjoy* science is to irrevocably prejudice your results. The idea of scientists as completely impartial curators of facts never sat easily with my understanding and experiences of science. It seemed antiquated, or robotic. At first during these epistemological disagreements, I would imagine my officemate as a Dr Who-style robotic scientist—a science-bot—sitting at a computer, running a laboratory in a mechanised, pre-programmed way. In response, I would argue that scientists give data meaning through our interpretations. I smugly suggested that if she really believed otherwise, she would submit a spreadsheet of raw, unprocessed and uninterpreted data to a journal for consideration and see how much luck she had with her 'objective' approach. In turn, she would respond with the ultimate disparaging retort, that I was merely a 'social scientist,' willing to make only vague, subjective statements of limited utility.

These feisty arguments with my officemate haven't yet been resolved. Both the 'science-bot' and the 'social scientist in disguise' are stubbornly resistant to being swayed by differing ideas about objectivity. An orthodox understanding of science as a system of knowing requires that science is an objective pursuit. There is no space for the subjective. Through this lens, my officemate's practice of science is arguably more valid than my own. However, such evaluations of validity of scientific practice are certainly not straightforward.

Let's return to 1834 for a moment, when a group of people seeking a name to describe their collective activities reached an agreement on the term 'scientist' (Anon 1834). That small but significant decision did not erase all humanity from the newly minted scientists. To my old officemate, it may seem like dangerous revelation that I am a person at all times, including when I do science. However, such a 'confession' is self-evident. Bruno Latour and Steve Woolgar's (1979) **tribe** of scientists contains a collection of individuals with cultures, beliefs and practices that produce scientific knowledge. I argue that the essential, inescapable humanity in all our pursuits, including our academic works, means that any associated research cannot be impartial and nor should it be.

Recognising the human influence on scientific knowledge is divisive in the broader scientific community. The protracted discussion between officemates reflects broader discussion in the wider scientific community, including within climate science, about the roles and responsibilities of scientists. Disagreements about the role of scientists, particularly in politicised fields of science, have been playing out within the climate science community for many years. Widely diverging opinions are regularly dredged up and turned over, before they are left to simmer for another day.

In 2013, for example, Dr Tamsin Edwards, a climate scientist at the University of Bristol, sparked another round in the recurring debate when she published a provocative opinion piece in the *Guardian* newspaper (Edwards 2013). In Edwards' (2013) piece, she implores her colleagues to remain scrupulously impartial. She argues for a single approach to doing climate science, one based on a unified 'moral obligation' to strive for impartiality and states quite definitively that climate scientists must not advocate for particular policies. She summarises her argument with the words of political scientist Robert T. Lackey:

> Often I hear or read in scientific discourse words such as degradation, improvement, good, and poor. Such value-laden words should not be used to convey scientific information because they imply a preferred...state [or] class of policy options...The appropriate science words are, for example, change, increase, or decrease.

Edwards (2013) op-ed sparked heated discussion[1]; many climate scientists saw her views as self-evident, but other scientists resolutely disagreed with her understanding of science. In particular, Edwards singled out prominent NASA Goddard Institute for Space Studies climate scientist Dr Gavin Schmidt,[2] who has conversely argued that climate scientists *should* state their policy preferences in order to avoid accusations of having vested interests or hidden agendas. The relationship between scientists and advocacy is concerning because there is little clarity and many opaque assumptions.

Schmidt (2014) has previously disagreed with other prominent scientists about the roles and responsibilities of climate scientists, as well as appropriate approaches to practicing science. For example, he argued in a *Nature Geoscience* commentary (2008) that blogs are a valuable resource within our community and also for communicating widely. They 'provide a rapid, casual, interactive and occasionally authoritative way of comment on current issues, new papers or old controversies' (Schmidt 2008, p. 208). Blogs have a capacity to help us 'engage, inspire and inform.' Meanwhile, Oxford University Professor Myles Allen (2008) disagreed. He argued that discussion must take place through the peer reviewed literature, our 'worst possible system...apart from all the alternatives...' (p. 209). In contrast to Schmidt, Allen recommends that any scientist who feels that they must 'communicate non-peer reviewed opinions to a journalist or member of the public, then stick to communicating one-to-one and make it clear you are speaking off the scientific record.'

My experiences of science, together with these broader community debates amongst climate scientists around responsibilities for communicating and addressing climate change, reveal a diversity of research practices. This spectrum of the objective to the subjective, or the **positivist** to the **constructionist,** is particularly contested in politically charged fields of science. In such heated disciplines, scientists typically question whether they should participate actively in the production of policy and decision-making processes, or whether doing so compromises an essential impartiality.

Professor Mike Hulme (2009) describes such complex, messy, labyrinthine types of research fields as **post-normal**, to distinguish them from typically **normal science** (Kuhn 1962). Normal science describes a period of discovery characterised by comparatively lacklustre research that occurs between explosive **paradigm shifts**. During this filling in period, facts are accumulated and progress is slow. We can think about normal science as guided by sociologist Robert Merton's (1973) four **classical norms** of scientific practice—**scepticism, universalism, communalism** and **disinterestedness**:

- Scepticism argues that all ideas must undergo rigorous testing and structured community scrutiny.
- Universalism requires claims of truth to be evaluated in terms of universal, objective criteria.
- Communalism describes the common ownership of scientific knowledge, by which scientists 'give up' their intellectual property in exchange for recognition and esteem within the community.
- Disinterestedness rewards scientists for acting in a manner that is not self-serving.

The suitability of these norms can be tested by returning to my fact-loving officemate, the hydrologist. She adopted highly analytical approaches to understanding environmental change in chemical systems. During her PhD examining the composition of various trace elements, she went out into the field, collected water and soil samples, returned to the laboratory and analysed these using an established set of techniques. Next, she collated her data and made interpretations about changes occurring in the river catchment and recommendations about watershed management. It's clear that her analytically focused discipline closer to a 'normal' science than the politicised minefield of climate science.

Conversely, in Hulme's alternative view of science, we characterise science as a method of inquiry where facts are inherently uncertain, values are disputed and the stakes are high. This is what philosophers of science Silvio Funtowicz and Jerry Ravetz (1991) term post-normal science. Here, we recognise gaps in knowledge and acknowledge messy perspectives. From a post-normal perspective, we no longer view climate change as a 'problem' that can be addressed with simple scientific 'answers,' but rather we see 'climate change [is] now as much as a societal problem as a physical one' (*Nature Climate Change* Editorial 2011, p. 1).

Climate change is a problem that demands more; it demands more *of* science and more *than* science. This broader context of climate change, the ideals of scepticism, universalism, communalism and disinterestedness are more problematic, restrictive, or at the very least, less useful guiding principles for climate science. Hulme argues that if scientists embrace climate science as a post-normal pursuit, we must accept that 'Where science is practiced, by whom and in what era, affects the knowledge that science produces. Science not only has a methodology, but it also has a history, a geography and a sociology' (p. 78).

Acknowledging the temporal and cultural context of science challenges a singular and static approach to science. How can we fold the history, geography and sociology of science into the idea of a universally appropriate and applicable way of being a scientist? Edwards (2013) describes climate scientists as bound together by a moral obligation to strive for impartiality. Such a universal plea prompts many intractable questions— who are scientists obligated to, and who decides a collective morality for science? Edwards is an accomplished and respected member of the scientific community. Her views (like everyone's) should be listened to and interrogated, and her opinion (like everyone's) is an important part of the debate on the role of scientists. But does such an opinion of objectivity constitute a moral benchmark that scientists must meet or be found lacking?

Edwards explains that part of the reason that she became a climate scientist was because she cares about the future of our natural world. This motivation fortifies her belief in the necessity for impartiality and objectivity in science to achieve the best outcomes for our natural world. Equally, part of the reason I became a climate scientist is because I care about the future of our natural world. This motivation fuels my belief in a reflexive, subjective and intimate connection to science to achieve the best outcomes for our natural world. If, for example, the manifestation of my care for the environment is a strongly held belief that we must invest in renewable energy as a matter of great priority, then I am necessarily immoral by Edwards' standards.

Edwards' call for a singular morality for all climate scientists reveals a diversity of scientific practice. We can glimpse a wonderful messiness in our human engagement with science in these differing manifestations of our care for the environment and our perceptions of morality. In my ignorance of social, economic and climate policy, I have little interest in advocating for particular social or economic policies. However, like Schmidt (2014), I adamantly believe that we are all advocates simply by being thinking, acting people. Scientists advocate for better funding, and scientists advocate for clear communication. Furthermore, scientists routinely defy Merton's (1973) norms of scientific practice by trumpeting our own self-importance and for the respect of our colleagues. While on the one hand scientists are encouraged to be disinterested and act in a manner that is not self-serving, in reality scientists are richly rewarded for promoting our own self-interests. Liberally advertising a recently published paper or throwing the necessary bombastic statements into a grant application are standard forms of self-advocating, and these practice essentially defy such a principle of disinterestedness.

Has acknowledging myself in my research, or more broadly the practice of science as contextual, harmed my credibility or the quality of my scientific research? I believe that my work has benefited from **reflexivity**. I strive to produce research that is **replicable**, that it is **rigorous**, that it is rooted in best practice and I expect that the conclusions of my research are **robust** to scrutiny. Nonetheless, my research is not conducted by a 'science-bot.' It is done by me—a living, thinking, complex human, whose identity, subjectivity, history and values are not magically erased or suppressed when I decide what research to undertake, how to interpret results or choose which journal to submit to. Denying the role of my human **subjectivity** in my science would be disingenuous. In opposition to Edwards (2013), I contend that scientist have *never* practiced science objectively; impartiality is simply not possible.

An Exclamation Mark

These differing understandings of the role of the scientist in science have implications beyond the individual practitioner or quibbles between stubborn scientists. I have argued that understanding science as a singularly objective and impartial pursuit inherently (and unnecessarily) negates the unavoidable role of the human. While this refutation is the *desired* outcome of the process-based method of inquiry embraced by my officemate in her hydrological studies, in the climate sciences this evidently denies an important element of both research practices and ultimately of the world, which I explore in more detail later. More broadly for all branches of science, any claim of objectivity is an untenable claim that distorts reality.

Returning briefly to the process of **Othering** I discussed in *Chapter 4*, my officemate's absolute rejection of our subjectivity is essentially an act of sanitising research that social scientist John Law (2004) describes as 'methodological hygiene.' Above all else, it is *subjectivity* that is concealed, that is cleansed from our scientific practices. I previously described my satisfaction at accomplishing the great intellectual marathon of my PhD, the scientific results of which could not readily be disentangled neatly from the web of my life and learning that had been woven around it. When I came to write and submit my thesis, these elements were neatly excised so that ultimately it was the subjective that was excised. This is typical, it is expected of us as individuals, yet we've already seen that this does not necessarily reflect how science is actually done. The rejection of the subjective, of the intuitive, of the human not only denies the reality of our ways of working and being, it denies elements of reality.

A rigid, unwavering rejection of the important of subjectivity in the production of scientific knowledge also obscures the boundaries of scientific inquiry. Climate change is a technically, conceptually and socially complex problem that is influenced by, and influences, humans (*Nature Climate Change* Editorial 2011). Such complexities must be acknowledged. Hulme (2009) asserts that it is not possible to see scientific knowledge, as Edwards does, as the neutral outcome of an objective pursuit and universal truth. Hulme argues that scientists must '…recognise and reflect upon their own values and upon the collective values of their colleagues. These values and world views continually seep into their activities as scientists and inflect the knowledge that is formed' (Hulme 2009, p. 79). Hence, classical principles of scientific inquiry do not need to be discarded, but simply revisited and revitalised. While *reaching* Merton's ideals is strictly untenable, Hulme encourages scientists to still *aspire* to these ideals, while also recognising and embracing the different flavours and inflections that infuse scientific knowledge.

An aspirational approach requires nuanced understandings of science and scientific practice. Thus far, I have deliberately placed my understandings of scientific practice in false opposition with my hydrologist officemate. While such arguments about science and its ways of knowing may feel intensely adversarial, they are not. These feisty arguments are not resolved by assigning one of us as right and the other wrong. Rather, these office squabbles and disciplinary spats reveal a diversity of practitioners ranging from 'science-bots' right through to 'social scientists' who have allowed themselves to be influenced by social theory or personal reflections.

I argue that rather than existing in opposition, there is instead a space between my officemate (the objective) and me (the subjective), a space beyond the objective as a fundamental scientific ideal where we openly acknowledge the culture, beliefs and plural practice of science. Borrowing (very loosely) from Law, I term this envisaged conceptual space the **hinterland**, in which we can place a more expansive set of descriptors of scientific practice.

For me, this conceptual hinterland was first inhabited by an exclamation mark. In 2011, I moved on from my exhausting PhD odyssey and started a new job researching extreme temperatures in Australia. I rapidly became very interested in both the technical and conceptual facets of my new research work. I was excited about applying new climatological methods to new contexts, and producing new, highly policy-relevant information. Equally, I had hesitations about elements of these state-of-the-art methodologies that I thought should be discussed openly and broadly, without necessarily detracting from the usefulness of our results. Such community-wide discussions seemed far beyond reach.

I remember flopping on the couch one winter evening in the early days of my new job. I absently flicked through a sociology textbook my girlfriend had discarded on the coffee table after a long day of her own PhD-ing. I was incredulous! The introduction was littered with personal anecdotes, experiences and thoughts, and more than that, the author had deployed a small army of exclamation marks in her quest to communicate with verve. I would have given anything to be able to use even a single exclamation mark in my professional undertakings! However, even a humble exclamation mark was an unrealisable dream to an 'objective, impartial' scientist.

As my long-running frustrations simmered, I eventually discovered blogging. While climatologist Myles Allen (2008) openly dismissed the utility of blogging, I found myself tumbling into Gavin Schmidt's (2008) camp of passionate bloggers. Like Schmidt (2008), I found that blogging had the capacity to 'engage, inspire and inform' me, in addition to a chance to reflect on my own values that inevitably infuse my work. Blogging gave me a space for these long craved for exclamation marks. It was an expressive medium—I could be frustrated, I could be thrilled, I could be doubtful. I chased this exclamation mark into the conceptual hinterland, striving for something that reflected my actual practice of science rather than concealing the most exciting or Law's 'dangerous' elements of research.

For me, this new space for scientific practice—this hinterland—wasn't just about simply shifting the conversation from the conference or university tearoom to social media. It wasn't simply just a new venue for new contributors. This exclamation mark revealed a fundamental change in the way I understood and practiced science. This reflected an opportunity for a more nuanced, transparent approach to my science.

A Shifting Contract

I have thus far only superficially described a conceptual hinterland, which I illustrate at first as occupied by an exclamation mark. The hinterland coined by Law is clearly a more expansive and ontologically intricate concept than intended by my appropriation of his notion. His hinterland emerges from Latour and Woolgar's (1979) observations of scientific practice at the Salk Institute, from which Law describes an elusive metaphorical hinterland as a background of already composed realities from which we can build further realities. Unpacking Law's realities and further realities is unlikely to provide climates scientists with tangible insights into their lived experiences, so I consider these intricacies of definitions and meanings beyond the scope of this discussion. Instead, I use Law's own words 'This, then, is the most important point: it is the character of this hinterland and its practices that determines what it is to do science, or to practice a specific branch of science.'

By focusing on what it is to *do* climate science, we can explore what else resides in my borrowed terms, the hinterland. I propose that the hinterland is filled up not only by scientists, but also from the other direction, through the changing expectations of society demanded by its **contract** with each science. Throughout the prior chapters, I have detailed the lengthy process of 'becoming' a scientist, from my earliest childhood memories of exploration of the natural world, to my formal university lectures in science and its ways of knowing and my frustrations at the systematic, expected **Othering**[3] of me from my research. Even within the few years since I completed PhD, the space in which climate scientists work—our contract—has shifted. The hinterland has opened up, ready to be filled by our implicit contract with society. Climate science must shift in return.

In the post-normal leaning fields of science, such as climate science, the necessity of a conceptual hinterland is revealed in broad desires to understand how scientists interact with science. It always surprises the cynical side of me that when I give public talks, the most common question I get asked isn't about whether climate change is happening. More typically, people are curious about how I *feel* about climate change and I am often asked about my anxieties about the future. Unsurprisingly, non-scientists are rarely interested in the nitty-gritty technical details or the fundamental concepts. Many of the scientific concepts that we use to talk about possible future climatic change are just that—complex scientific concepts.

For example, climate scientists are concerned with pinning down a number for the Equilibrium Climate Sensitivity (ECS) (Knutti and Hegerl 2008; Otto et al. 2013). The specific details of ECS are unnecessary for this discussion, but put simply this is the global average surface warming response to a doubling of atmospheric carbon dioxide concentrations after the system has settled and reached a new steady state. In the real world, it could take hundreds of years or longer to see what temperature change would result from a doubling of carbon dioxide levels in the atmosphere. This ECS value is particularly tricky to establish in scientific analyses, so climate scientists use a combination of approaches, from both observations and climate models, to determine the range of possible sensitivity values.

Understanding climate sensitivity is undeniably important for a scientific understanding climatic process and change, but how much does the exact ECS range matter to people? Does it matter to you if the temperature response in 500 years to a hypothetical doubling of carbon dioxide is precisely 3.5°C instead of 4°C, or are there other, more pressing concerns for society?

Much of my recent work has focused on recent temperature extremes in Australia, which keeps my wider research group very busy. In the last decade, Australians have endured various extremes—record 2009 bushfires and temperature, record 2010 and 2011 flooding, and from spring 2012 onwards, extreme heat dominated our weather and climate (Bureau of Meteorology 2012, 2014). In 2013 alone, Australia experienced its hottest day, week, month, season and year, in an observational record extending back to 1910. These events became the focus of an ongoing research programme for our group. Can science provide insight into the physical processes behind these extremes? Can science determine how these events relate to anthropogenic climate change or natural climate variations? Can science evaluate, characterise and quantify the physical world?

While our research team was highly successful in investigating these extreme events from a scientific perspective, these scientific questions are just one way to relate to extremes. If scientists analyse an observed extreme and ultimately calculate that the risk of a particular event occurring was 30 times more likely due to anthropogenic influences (Lewis and Karoly 2014), what then? These extreme climate events have important meanings beyond 'objective' scientific analysis and results.

I regularly present my scientific results at public lectures or community events. I used to show a photo depicting a Tasmanian family sheltering under a pier from a fire front. The sky is suffused with fire. In the ocean, a grandmother holds two children while their sister helps her brother cling to underside of the pier. After a few talks, I had to remove the photo from my PowerPoint presentation because each time I turned around to discuss it, it would make me teary. Later, I talked about this experience of connecting so viscerally with that family during a Vox Pop that was filmed by a climate change communication group. After an afternoon of shooting, I buried myself back into a pile of research tasks and forgot about the video until I started getting emails from strangers encouraging me to stay strong, or that it was quite okay for a scientist to feel.

When we filmed that footage back in 2013, I felt keenly that the year we were living was a chilling taste of our world to come. Just outside of Sydney, tinderbox conditions occurred in early spring of 2013, following a dry, warm winter. Bushfires raged far too early in the season. Further south in the state of Victoria, a higher than average bushfire risk warning was issued for the forthcoming summer. This is what such extremes mean; record temperatures have serious impacts. At that time, I was frightened by what's to come in a world 1 °C hotter than now, regardless of what the ECS turns out to be. At public lectures and community events, people want to know that I am frightened about bushfires. They want to know that I am concerned about the vulnerability of our elderly to increasing summer heat stress. People want to know that, amongst everything else, I remain optimistic about our collective resilience and desire to care for each other.

There is a hinterland beyond our efforts to evaluate, characterise and quantify climate change and variability. Public questions about how I feel, or kindly but unasked for emails reassuring me that it is only human to worry for victims of natural disasters, are a way in which society invites scientists to acknowledge and to inhabit the hinterland. This shifting desire to re-negotiate our contract with society is evident in the resonance of projects such as *Is this how you feel?* (Duggan 2005), in which prominent climate scientists are asked to share their feelings about climate change in hand written letters. These letters are 'the words of real climate scientists.'

Like me, these scientists experience an amalgamation of fear and hope. They worry for their only-dreamed-of children, or for their existing grandchildren. They are angry. They are sad. They are positive. They are apologetic. They are excited by their work. They are ashamed of their professional enthusiasm. Dr Ailie Gallant, climate scientist at Monash University typifies such feelings:

> I feel nervous. I get worried and anxious, but also a little curious. The curiosity is a strange, paradoxical feeling that I sometimes feel guilty about. After all, this is the future of the people I love.

Many of these letters have resonated with the public, sparking conversations along the lines of, 'if climate scientists are worried, maybe I should be too.' These letters inhabit my proposed hinterland.

These emotional responses to suffering and connections to society they do not betray Merton's norms of scepticism, universalism, communalism and disinterestedness, and nor do they weaken our scientific claims. Rather, these enactments of science centred firmly on the practitioner (the scientist) are a necessary and useful response to the limitations of normal science. These letters, kindly emails and my own desired exclamation mark are an acknowledgement that 'if scientists want to remain listened to, to bear influence on policy, they must recognise the social limits of their truth seeking and reveal fully the values and beliefs they bring to their scientific activity' (Hulme 2013, p. 88). As a corollary, this acknowledgement of limits requires a willingness of science to reappraise its contract with society and shift in response.

THRESHOLD MOMENTS

The obligations of a mutable contract are not discharged simply by scientists providing more, or different, scientific knowledge to society. Acknowledging a changing broader context for science also requires changing the norms of scientific practice, and requires scientists collectively to adapt in how we behave, and who we are as scientists. We can explore the necessity for this change through the example of issues brought to light by the 'climategate' emails (Castree 2013; Mann 2013). In a calculated attack on the legitimacy of climate science that occurred just before an important round of climate policy negotiations in Copenhagen in November 2009, computer servers at the University of East Anglia were illegally hacked and email correspondence was stolen.

A selection of these emails between climate science colleagues was published on the Internet, which focused on quotes that purported to reveal dishonest practices that promoted the myth of global warming. For example, the scientists discussed amongst themselves several 'tricks' with their data and their disappointment at the 'lack of warming' this 'decline' would require hiding. Several independent inquiries subsequently investigated the emails and it turns out that each of these scandalous revelations had a scientific basis. The 'tricks' were shorthand for a new statistical technique and the 'decline' referred to physiological changes in tree ring width (a proxy for climatic change) that occurs as trees age. The climategate scientists were exhaustively cleared of wrongdoing. Their research practices and scientific findings were found unimpeachable, even under the most intense of scrutiny.

On the surface, the climategate emails were an unpleasant but unremarkable event. However, delving a little deeper, this can be seen as a significant turning point—a **threshold moment**—in society's relationship to climate change. Within the political and social space, the leaking of the emails conspired with several key factors. The political negotiations at Copenhagen in 2009 had been vaulted as crucial talks that would fundamentally redress our global sluggishness in combating climate change. But the talks then failed. In addition, the Earth experienced slower rising global surface temperatures than in the previous decade, which sparked a persistent sceptic trope that global warming had ended (*Nature Geoscience* Editorial 2014).

All in all, the momentum of the previous US Vice-President Al Gore-filled years that was driving changes in our collective understandings of climate change faltered. Even now, some years later, the impact of the release of the stolen climate emails still persists. In the years since, the discourse between climate change deniers and advocates has degraded to previously long-dead arguments about the fundamental physics and the reliability of data and physics. These déjà vu arguments still dominate the discourse around the science of climate change.

The circumstances around the leaked climate emails represent a threshold moment in society's expectations of scientists. On one hand, these were dedicated, meticulous and renowned scientists who were conducting important research and in doing so, they unwittingly became the subject of an insidious and illegal attack, perpetrated with the intention of undermining their credibility and that of all climate scientists. The leaking of the emails very sadly ruined ordinary lives.

Another parallel interpretation is that the content of the emails, and the response of scientists and their institutions to the leaking of the emails, reveals a rift opening into our hinterland. The climate scientists and their institutions rejected wrongdoing and the numerous fastidious reviews of their conduct support their claims. However, the persistent dissatisfaction with the outcomes of the leaking and the response demonstrate a societal 'need' of scientists that is not being met with this response, which characterises 'normal,' rather than 'post-normal' scientific practices.

These two readings of the climategate email scandal—that the scientists were faultless and that the circumstances reveal an inadequacy in scientific practice—are not mutually exclusive. These reading do not even sit side-by-side. I do not suggest the scientists who sent and received these emails, behaved illegally or immorally or even dubiously. Rather this particular event, this turning point in the discourse, is layered with various meanings. These stolen emails represent a threshold moment.

This threshold moment lies outside our current framework of research practice and research integrity, but within the framework expected of us by society. We can see this confluence of events and timing and imbalance between responses and expectations as a reflection of the inevitable flexing and re-settling of science's contract with society. In further instances, scientists have clashed heavily with climate sceptics who demand access to data, computer scripts or emails. These requests are often intended as a form of harassment, and scientists typically push back, with good justification (Lewandowsky and Bishop 2016). A cultural unwillingness to share scientific information with sceptics is well within the bounds of our traditional expectations of good scientific practice and does not inherently reflect that there is something being hidden away and obscured from view. Although this rejection of sharing information publicly fits soundly within the self-described and self-adjudicated ideas of scientific research integrity, it no longer encompasses what society expects. That is, there are other facets of research practice that the public expects, which are not yet met by scientists.

The desire for public connection with the processes of science and the outcomes of scientific pursuits is not new. In the late 1700s, scientific experiments were a form of entertainment, and scientific lectures were a hugely popular source of entertainment for Europe's elite. The very latest discoveries were demonstrated in theatrical style. Over time, trained scientists usurped the self-funded gentleman scholar, and science was steadily secreted away, into the relatively modern construct of the laboratory.

Being a scientist became synonymous with knowing one's way around a laboratory (Rouse 1987), and as a result, the production of scientific knowledge became increasingly specialist and increasingly inaccessible.

In a return to days passed, the contemporary **academy** is more and more expected to be a public institution, engaging and communicating openly with the public on their terms. As revealed through the climategate email event, the public is no longer necessarily satisfied by the learned people accomplishing learned knowledge in learned institutions. In particular, the knowledge obtained through this process are typically secreted away and obscured by pay walls,[4] which cannot be publicly accessed. That is, a great transparency of science is required. Hence, to add to my long-coveted exclamation mark, we can now fill our hinterland with the idea of scientific transparency.

Scientists Chris Rapley and Kris De Meyer (2014) also identify this gap between the current role of the climate science community and the needs of society. Rapley and De Meyer attempt to question how climate scientists should balance their efforts between research and engagement, and whether, like Edwards, they should aim to strictly inform, or instead they should advocate for specific actions. The authors do not explicitly attempt to resolve these questions but rather they 'encourage the community to reconsider its professional practices, skills and norms, and to adjust its training and development activities accordingly' (p. 749). That is, science cannot continue to maintain Allen's (2008) dismissal of communication of science outside peer reviewed literature, which obscures the processes of science and its knowledge outcomes. Rather, scientists *must* acknowledge the meaning of our threshold moments and their significance for revealing this conceptual hinterland beyond the objective. This hinterland is a nuanced place, where the public desire for transparency is recognised and acknowledged as legitimate, important and not intrinsically intrusive.

We won't all be packing a bag for the hinterland. Just as I understood in *Chapter 4* that my PhD-writing frustrations would not be ultimately resolved by a generation of young bright-eyed climate scientists presenting auto-ethnographies as part of their dissertations, I understand that scientific will not all lurch towards a nebulous hinterland. It is clear that scientist will not and should not all respond to the threshold moment of the climategate emails by indiscriminately discarding the practices considered necessary or useful to the practice of science. That is, climate scientists will not strive for the idea of transparency at the expense of traditional respect for data ownership and publication. Furthermore, it is clear that scientist will not all respond to my individual frustrations of experiencing the subjective in my research practices and striving to breach this hinterland.

Not all individual scientists will reveal their feelings publicly about the challenges of climate change, and nor should they. At times it seems incredible that my loose, wishy-washy ideas about science and its way of knowing can cohabit a name with those who champion impartiality and fact collecting, those affectionately described 'science-bots' such as my officemate. Nonetheless, this diversity of opinions on issues as fundamental as who we are and what we do is a great strength of science. Scientists can be science-bots, can believe in striving for impartiality, can place great trust in the relevancy of scientific facts or scientists can allow themselves to be influenced by social theory or personal reflections. Scientists do not all need to inhabit this hinterland beyond the objective, but collectively, scientists must recognise its existence and utility.

Notes

1. The irony, of course, is that in penning her opinion piece, she contradicted her argument that scientists should not express personal opinions.
2. In the spirit of transparency, I should note here that I have worked closely with Gavin and we have been jogging a few times.
3. A more detailed discussion of Othering is provided in *Chapter 4.*
4. See *Chapter 5* for further information on academic publishing.

Glossary

Academy   The academy is the institution and community concerned with the pursuit of research and scholarship.

Classical norms   These are Merton's (1973) principles of good scientific practice (scepticism, universalism, communalism, disinterestedness).

Communalism   Communalism is one of Merton's (1973) norms. This describes the common ownership of scientific knowledge, by which scientists 'give up' their intellectual property in exchange for recognition and esteem within the community.

Constructionist   The concept that humans generate knowledge and meaning from an interaction between their experiences and their ideas.

Contract A concept of Hulme's (2009) to describe the critical relationship between science and society, which is bound by a contract of understanding and obligation.

Disinterestedness Disinterestedness is one of Merton's (1973) norms. Disinterestedness rewards scientists for acting in a manner that is not self-serving.

Hinterland This is a term coined by Law (2004) to describe a background to reality that we can build further realities. The hinterland describes a geography of reality—a topography of reality possibilities. Here, I use the term to describe a space beyond positivist approaches that permit new narratives for science.

Normal science Following Kuhn's (1962) view of scientific knowledge creation, normal science is the period of discovery between paradigm shifts, where facts are accumulated and progress is slow.

Objective/objectivity The ideal of the objective in science is the idea that scientific claims and results are true outside of, and not influenced by, a scientist's individual biases, interpretations and perspectives.

Othering The process by which researchers exclude information that does not fit into an approach as irrelevant.

Paradigm shift In Kuhn's (1962) view, a scientific revolution (or paradigm shift) occurs when scientists encounter anomalies which cannot be explained by the currently accepted paradigm. A paradigm encompasses the entire consensus worldview of scientists. Following a shift in paradigms, scientists do not reject the subsequent paradigm as redundant, as this can remain useful though limited. A classic example is the transition between Newtonian physics and Einsteinian relativity.

Positivist The concept that information derived from sensory experience and interpreted through rational and logical approaches is the true source of authoritative knowledge. Intuitive knowledge is rejected in favour of empirically based knowledge.

Post-normal science Philosophers of science Funtowicz and Ravetz (1991) describe science as post-normal where we recognise it as a method of inquiry where facts are uncertain, values are disputed and the stakes are high.

Replicable Research that is capable of being duplicated or repeated using the same approach.

Rigorous I use this broadly to describe research that is trustworthy and utilises the appropriate tools and approaches.

Robust   Again, this is a broad term to describe research that withstands the firmest scrutiny.

Reflexivity   Here I refer to a form of personal reflexivity, where I acknowledge that my values, beliefs and interests have influenced by research.

Transparency   An openness in communication and accountability of scientific practice such that non-scientists can view how scientific knowledge is generated.

Tribe   Latour and Woolgar's (1979) description of scientists as bound together by a set of practices. This is summarised by Law (2004) as 'Scientists have a culture. They have beliefs. They have practices. They work, they gossip, and they worry about the future. And, somehow or other, out of their work, their practices and their beliefs, they produce knowledge, scientific knowledge, accounts of reality' (p. 19).

Scepticism   Scepticism is one of Merton's (1973) norms. This encapsulates the ideal that all ideas must undergo rigorous testing and structured community scrutiny.

Subjective/subjectivity   This contrasts with objectivity and refers to the idea that a scientist's judgements are necessarily shaped by their personal opinions and feelings.

Universalism   Universalism is one of Merton's (1973) norms. This requires claims of truth to be evaluated in terms of universal, objective criteria.

References

Allen M (2008) Minority report. Nature Geoscience 1:209–209. doi: doi:10.1038/ngeo174.

Anon (1834) On the connexion of the physical sciences. The Quarterly Review 51:54–59.

Bureau of Meteorology (2012) Australia's wettest two-year period on record; 2010–2011. Special Climate Statement 38. National Climate Centre Bureau of Meteorology, Melbourne.

Bureau of Meteorology (2014) Annual Climate Report 2013. National Climate Centre Bureau of Meteorology, Melbourne.

Castree N (2013) Making Sense of Nature. Routledge, New York, NY.

Duggan J (2015) Is this how you feel? In: isthishowyoufeel.weebly.com.http:// isthishowyoufeel.weebly.com/this-is-how-scientists-feel.html. Accessed 13 Jun 2015.

Editorial (2011) Whole-system science. Nature Climate Change 1:1–1. doi: 10.1038/nclimate1083.

Editorial (2014) Hiatus in context. Nature Geoscience 7:157–157. doi: 10.1038/ngeo2116.

Edwards T (2013) Climate scientists must not advocate particular policies. In: The Guardian. http://www.theguardian.com/science/political-science/2013/jul/31/climate-scientists-policies. Accessed 8 Apr 2015

Funtowicz SO, Ravetz JR (1991) A new scientific methodology for global environmental issues. In: R Costanaza (ed) Ecological Economics the Science and Management of Sustainability. Columbia University Press: New York, NY, USA, pp 137–152.

Hulme M (2009) Why We Disagree about Climate Change: Understanding Controversy, Inaction and Opportunity. Cambridge University Press, New York.

Hulme M (2013) Exploring Climate Change Through Science and in Society: An Anthology of Mike Hulme's Essays, Interviews and Speeches. Routledge, London.

Knutti R, Hegerl GC (2008) The equilibrium sensitivity of the Earth's temperature to radiation changes. Nature Geoscience 1:735–743.

Kuhn TS (1962) The Structure of Scientific Revolutions. University of Chicago Press, Chicago.

Latour B, Woolgar S (1979) Laboratory Life. The Social Construction of Scientific Facts. Sage Publications, Beverly Hills.

Law J (2004) After Method: Mess in Social Science Research. Routledge, London.

Lewandowsky S, Bishop D (2016) Don't let transparency damage science. Nature 529:459–461. doi: doi:10.1038/529459a.

Lewis SC, Karoly DJ (2014) The role of anthropogenic forcing in the record 2013 Australia-wide annual and spring temperatures [in "Explaining Extremes of 2013 from a Climate Perspective"]. Bulletin of the American Meteorological Society 95:S31–S34.

Mann ME (2013) The Hockey Stick and the Climate Wars. Columbia University Press, New York.

Merton RK (1973) The Sociology of Science: Theoretical and Empirical Investigations. University of Chicago Press, Chicago.

Otto A, Otto FEL, Boucher O., et al (2013) Energy budget constraints on climate response. Nature Geoscience 6:415–416. doi: 10.1038/ngeo1836.

Rapley C, De Meyer K (2014) Climate science reconsidered. Nature Climate Change 4:745–746. doi: 10.1038/nclimate2352.

Rouse J (1987) Knowledge and Power: Toward a Political Philosophy of Science. Cornell University Press, Ithaca.

Schmidt G (2008) To blog or not to blog?. Nature Geoscience 1:208–208. doi: 10.1038/ngeo170.

Schmidt GA (2014) On scientists and advocacy. Science 344:256–256. doi: 10.1126/science.1251819.

CHAPTER 7

Blue Skies and Other Shades

Abstract Lewis explores the importance of curiosity as a key aspect of scientific practice. By exploring a case study of palaeoclimate research and examples of the dark outcomes of scientific research, Lewis demonstrates the limitations of a utilitarian framing of research. Such a problem-solution approach to scientific inquiry denies the essence of science as an experimental, uncertain and curious mode of producing knowledge. Lewis argues that scientists should also view science as a fundamental creative process that requires curiosity. Hence, curiosity and transparency are presented as counterpoints in scientific practice. By employing both as key elements of scientific practice, scientists can both attend to the critical relationship between science and society, and develop a deeper and richer connection to the world.

Keywords Curiosity · Transparency · Palaeoclimate · Value · Useless knowledge · Problems

I next add **curiosity** as a new key descriptor for scientific practice. To discuss the necessity of curiosity, I first unpack the idea of **problems**, arguing that understanding research as simply finding solutions to easily defined problems is in itself problematic. Using both a case study from my own research experiences of studying palaeoclimatology and examples of the dark outcomes of scientific research, I demonstrate the limitations of

© The Author(s) 2017
S.C. Lewis, *A Changing Climate for Science*,
DOI 10.1007/978-3-319-54265-2_7

framing either individual studies or entire fields of research as the solutions to specific problems. Such a utilitarian framing denies the essence of science as an experimental, uncertain and curious mode of producing knowledge.

By investing only in providing solutions to problems, science risks erasing vast swathes of interesting research that is not produced in response to an identified 'problem.' Scientists should also view science as a fundamental creative process that requires curiosity. I reframe problems as not as difficulties and complications, but as a necessary part of a creative process that allows both scientists to attend to a deeper and richer connection to science, and allows science a greater attention to the world. This revisiting of the idea of problems requires a commitment of scientists to curiosity.

This Study Will Solve Everything

A few of years ago, my girlfriend was attending a major sociology conference with a small group of her PhD friends. Listening to them talk each evening as they settled in with a hard-earned beer to debrief from a long day of listening and networking, it seemed that they were both amused and dismayed by various conference papers. These papers highlighted various social issues and then problematised these issues in a way that was, well, problematic. The takeaway message from the conference seemed to be that sexting, selfies and teenage binge drinking all herald the demise of society. From what I could infer, this narrow approach to sociological research starts by identifying an issue as an inherently harmful social problem that must be discussed, analysed and ultimately solved through research.

The group of PhD students was disappointed at this functional view of sociological research. While a core tenant of the social science is to question what others might take for granted, the approach communicated in many of the conference talks conversely re-inscribed popular perceptions. Such a functional approach also transpires in scientific research, where questions are also framed around the idea of problems that need to be solved. It may seem tangential to begin by unpacking the ideas of problems and solutions—the usefulness of this approach seems self-evident. However, in my experience, such problem-solution approach to academic inquiry is not always the most insightful nor entirely sufficient.

Around the same time the annual national sociology conference was taking place, I attended a workshop for early career researchers. Workshop participants were tasked with bringing along a research paper was near completion or at some stage of the peer review process. I turned up with the key findings of my research summarised and was given tips on how to pitch my research for a general audience so that I would be ready to be interviewed by a hypothetical journalist about my new results. Next, the group practiced communicating the implications of our studies and why the public should care about our research endeavours. We were given some excellent practical advice—I came up with some interesting analogies to translate my technical findings into more readily understandable concepts and finally, by the end of the day, I was able to explain why my research was important in clear and simple language.

While pleased with my newly honed communication skills, I was hesitant about how this approach framed scientific inquiry. The research paper I came prepared to discuss really wasn't particularly interesting (see Lewis et al. 2014 for yourself). By the time I had followed the streamlined communication approach and taken the leaps from my results findings to the easily interpretable findings and implications that were digestible by a general audience, I had joined that particular group of gently derided social scientists who equated sexting with the disintegration of the world as we know it.

My paper was esoteric and technical and reached frustratingly vague conclusions. In short, it had nothing definitive to offer the reader. My study used new approaches to look at old research questions and was deemed novel enough to pass through the scrutiny of three peer reviewers. I thought it was a useful analysis, but it certainly didn't seem all that interesting. Using the techniques I learned in the communications workshop, this single, dull paper seemed like it could save the human race. The rapid transformation of my research paper from Clarke Kent to Superman concerned me.

First, when *every* paper is discussed as ground-breaking, it diminishes the spotlight on those that are actually revolutionary and necessary of wide dissemination and discussion. When every paper is presented as *the* study that will solve everything, it necessarily becomes harder for expert and non-expert alike to understand what is important, and why and how it fits into the wider body of scientific research. Indeed, it can become difficult to trust any results when scientists insist that every new study, for example, reveals the ultimate cure for cancer or weight loss. This creates a pervasive impression that only bombastic research matters. Second, when papers are invariably framed as 'solutions' to 'problems,' scientists are conducting a very reductive type of science, which denies vital nuance in understanding the multiple facets and complexities of the world.

Satisfying the Mums and Dads

Writing in Harpers magazine Abraham Flexner (1939) extolled the usefulness of **useless knowledge**. Wondering whether the tendency towards a utilitarian approach to social, economic and governmental issues had become too pervasive and powerful, Flexner suggests that this functional approach to research leaves too little room for opportunity—what is considered **useful** has 'become too narrow to be adequate to the roaming and capricious possibilities of the human spirit' (p. 544). Flexner focuses on the limitations of usefulness in a narrow sense, meaning practical and applied research, and argues that at the time this very definition of useful had had come to exclude 'the overwhelming importance of spiritual and intellectual freedom' in research.

In his article, Flexner raises important issues around how society assesses the usefulness of research and subsequent knowledge. He first notes that what is useful can only be determined retrospectively, sometimes many years or decades after such a discovery is made. Flexner provides examples of work that sat quietly unnoticed in a corner for long periods of time before it became valuable, including the discovery of the Ehrlich technique as a demonstration of latent usefulness. He reminds us of the great extent to which the 'pursuit of useless satisfactions proves unexpectedly the source from which un-dreamed-of utility is derived' (p. 544).

In 1870, for example, Paul Ehrlich was a very young student at the University of Strasbourg. Prone to highly focused, self-contained research, Ehrlich eventually graduated and went on to Breslau, which was then in Germany, where he continued his solitary style of work. Ehrlich described this as 'Ich probiere,' or roughly translated as 'just fooling,' though this would now be termed **blue-sky research**. While Ehrlich was busy 'fooling' at the boundaries of the emerging field of bacteriology, a Breslau colleague went ahead and applied Ehrlich's staining techniques to bacteria. The applications of this 'fooling' are now used to examine human blood samples for the purpose of differentiating types of bacteria. The

Gram staining process remains a contemporary scientific practice, still used in thousands of hospitals worldwide every day.

Flexner provides further examples of valuable results emerging from such free spirited wanderings or seemingly unconnected endeavours, including application of the abstract Non-Euclidean Geometry that allowed the development of the theory of relativity, the application of chemistry to the development of weaponry and of Maxwell's calculations of magnetism and electricity that were used by Marconi in developing the radio. These seeds of great discoveries had no practical objectives, but were prompted instead by a desire to satisfy curiosity.

Nearly 75 years after Flexner's entreaty to free the human spirit through academic inquiry, the concepts of useful and futile research are still widely discussed in Australia. On the brink of the September 2013 Australian federal election, one major party vying to hold government emphatically promised to target 'futile' research as an opportunity to apply funding cuts. Various electoral candidates put forward several examples of useless research, particularly in the humanities and arts, including one research project 'Spatial Dialogues: Public Art and Climate Change,' which received funding from Australia's peak funding body, the Australian Research Council (ARC).

Research bodies and individual researchers responded by highlighting the critical need for independence in allocating research funding and evaluating outcomes. The outcry at promised cuts also prompted the compilation of lists of supposedly futile Australian research that turned out to be conversely far from worthless, in the same manner as Flexner's examples of great discoveries that sat latent for many years. The most obvious recent Australia example is CSIRO scientist Dr John O'Sullivan et al.'s (1996) search for exploding black holes that led to his discovery of the now ubiquitous Wi-Fi technology.

Academic research is often targeted in this way, with newspaper articles deriding research a tiresome recurring event. More recently, Australia academics argued their worth when journalists suggested that the ARC should be forced to 'justify its grants in the front bar of a pub in western Sydney or northside Brisbane' (see Lamberts and Grant 2016). In November 2016, Australian Minister for Education and Training, Senator Simon Birmingham, concurred that academic research should 'pay dividends for Australian young people, old people, mums and dads.' A cursory Google search reveals that such criticisms of research funding and prescriptive directions for research outcomes are almost an annual event, spanning multiple countries, with the ARC, the USA's National Science Foundation and UK research councils all publicly ridiculed for purportedly funding nonsense work with taxpayer money (Mervis 2014a, b).

Dr Rod Lamberts and Dr Will Grant (2016) from the Centre for the Public of Awareness of Science at the Australian National University describe these routine denunciations of research as 'lazy swipes by lazy blowhards at lazy academics lazing their way through granting procedures (notwithstanding the fact that these procedures are hyper-competitive).' Echoing Flexner, Lamberts and Grant argue that such swipes are fallacious for several reasons, including that we don't know what research is valuable and should be funded, '...it's impossible to tell which individual idea or piece of research might trigger the next revolutionary breakthrough.'

A Case Study

The complexities of considering the value and utility of academic inquiry can be revealed through examining a case study of my research. What about my PhD research field? Is this work of intrinsic value, and it might one day prompt a breakthrough to rival Ehrlich's? Do the outcomes of this research satisfy the yardstick of Australia's mums and dads? Or is this instead the work of a lazy academic lazying through my research on the taxpayer's purse?

In *Chapter 4* I discussed my experiences of writing my PhD thesis on palaeoclimatology, the study of past climates. Was this useless scientific research? If asked questions of value and utility, palaeoclimatologists typically provide an enthusiastic answer centred on the idea that understanding past climates is necessary for understanding contemporary and future climate change. The instrumental climate record (e.g. from thermometers or rain gauges) is too short to reveal the range of natural climatic variability occurring on timescales beyond a few years or decades. Beyond these instrumental records, knowledge of past climates is derived from proxy records, which are natural archives such as ice cores and tree rings that preserve climatic information such as temperature and rainfall changes.

Enthusiastic palaeoclimatologists will likely proclaim that palaeoclimatic reconstructions provide a valuable opportunity to understand the range of natural climatic variability and potentially a context for understanding projections of future climatic change. This is an attractive idea;

permeating graduate students' thesis proposals and lucrative research grant applications alike. I have framed my own research of past climates in these terms, invoking this justification for my academic work. During the later stages of my PhD, waking during the night in a cold sweat of anxiety, I was convinced that my project was necessary for safeguarding the water future of vulnerable people in a warming world. At this crucial stage of my research, a sick day, a sleep-in or a holiday was utterly unthinkable. My research was simply too important.

But was it really the case that my study of past climates was useful because it shed a brighter light on possible future climate changes? Such an assessment is complicated. First, there are several ways in which palaeo-climate research enhances our understanding of the climate system, and provides useful information about contemporary change. The following examples are not exhaustive, but provide a snapshot of this type of insightful research:

1. *Constraining climate sensitivity.* Climate sensitivity is a measure of how responsive the climate system is to a particular **radiative forcing** (Knutti and Hegerl 2008), providing an assessment of the change in temperature that occurs from giving the climate system a kick from an external radiative forcing (e.g. adding carbon dioxide (CO_2) to the atmosphere). Climate scientists are particularly interested in **equilibrium climate sensitivity**[1]. (ECS). An important characteristic of the climate system, the ECS refers to the long-term change in the global average surface air temperature following a doubling of carbon dioxide in the atmosphere. I think of this how angry the earth's system will be if we whack it on the head with a sledgehammer filled with long-lived greenhouse gases and wait for the ensuing lump to abate.

   The ECS is thought to be in the range of 2–4°C (IPCC 2013), although estimates of this value are expensive to obtain using computer models. They require very long climate model experiments of many thousands of model years, so that the slow processes that occur in the ocean have time to play out and impact the other components of the Earth's system. A study by Andreas Schmittner and his colleagues (2011) tried to narrow down the ECS using palaeoclimate approaches. They combined temperature reconstructions from the height of the last ice age (the Last Glacial Maximum (the LGM)) with climate model simulations and determined that a doubling of CO_2 would lead to a global warming of between 1.7 and 2.6°C. Overall, Schmittner's approach combining various techniques produced a climate sensitivity value that was a lower than previously estimated using model experiments. While technical elements of the study were critiqued (Fyke and Eby 2012), it demonstrated the role that palaeo-based studies can have in investigating important characteristics of the climate system.

2. *Evaluating global climate models.* Global climate models are routinely evaluated to determine their skill in recreating important elements of the observed climate system. These measures of skill can be anything from whether a model captures the observed inter-annual variability of the climate system, or the seasonality of precipitation in a particular location, or the response of the model to a known volcanic eruption. However, these metrics, or measures of skill, are limited and do not cover everything climate scientists need to know about how well a model performs.

For example, these metrics do not push the models as much as what climate scientists expect to occur during the twenty-first century and hence greater insight into model performance can be obtained by using model simulations of past climatic change, when the system was substantially different from now. The Paleoclimate Modeling Intercomparison Project (Braconnot et al. 2011) is a coordinated endeavour involving modelling groups from different institutions across the world. Groups participating in this project each perform standard, agree upon experiments that allow the various climate models to be compared to each other. The palaeoclimate modelling experiments include simulations of the climate of the last millennium (1,000 years), of approximately 6,000 years ago (the mid-Holocene) and of the LGM.

The PMIP effort occurs alongside the Coupled Model Intercomparison Project (CMIP5) (Taylor et al. 2012), which is a larger suite of experiments that investigate the climate of the twentieth century through to scenarios of future climate change. The same models participate in both projects and allow us to compare the palaeoclimate simulations with proxy reconstructions during times where significant climatic changes occurred and provide important information for understanding future changes (Schmidt 2010).

3. *Providing a long-term context for detection and attribution studies.*
An extensive research project recently brought together climatologists, palaeoclimatologists, natural resource managers and historians, in order to reconstruct Australia's climate history. This project linked together the instrumental climate record, with historical documents and proxy reconstructions, and described the impact of climate variability on Australian society (Gergis and Ashcroft 2013; Fenby and Gergis 2013). Similar projects have been undertaken in other regions around the world and have assisted in **detection** and **attribution** studies, which aim to establish the most likely causes of climatic change. Detection and attribution studies, like most statistical analyses, require large datasets. The longer the climate records that are available for investigation, the more statistically confident scientists can be about both detected changes and their underlying causes.

   This multi-disciplinary Australian climate history project focused on reconstructing temperature changes over Australia using coral, tree rings and stalagmite records for the last thousand years. Using this approach, the researchers investigated the decadal drought in southeastern Australian from around 1997 to 2009 in a longer-term context than would be possible using the instrumental record (Gergis et al. 2011). The authors determined that this 'Big Dry' was likely to have been the worst since European settlement and that there was a 97% chance that this particular drought decade was the driest since 1788. This information is useful for understanding the risks and likelihood of recent extremes.

Arguably, these examples of palaeoclimatic research have informed scientists' understandings of the climate system in a tangible way. But what about research that does not necessarily shed light onto our uncertain future? If we can only draw a tenuous link between a project's research goals and our ability to understand what climatic changes lie in store next year or next decade, what then?

Returning to my earlier description of my own PhD, it is obvious that I exaggerated the importance of my work. Rather than protecting vulnerable people from catastrophic shifts in rainfall patterns in our warming world, my PhD research was rather arcane. I spent several years analysing the output of global climate models. This was not the more commonly understood process of examining climate models and making statements about likely future changes. Instead, I investigated changes during a specific type of climatic event, the **Heinrich events**[2], which occurred periodically during the past, with the most recent taking place around 15,000 years ago.

Furthermore, rather than investigating the dramatic impacts of these events on global climate, I actually examined the impact of these events on the global distribution of **water isotopes**, a chemical signature that serves as a useful tracer of hydrological change. My research looked specifically at how the location of rainfall to particular areas changed during abrupt Heinrich events when using model simulations that included isotopes (Lewis et al. 2010).

Overall, my thesis revealed that in simplified climate model experiments that replicate Heinrich events, the source of rainfall to East Africa might be more dominated by the Atlantic rather than Indian Ocean. My scientific results are hardly comparable in value to Ehrlich's fooling that stumbled upon an important method for bacterial staining.

50 SHADES OF SCIENCE

My confession of the insignificance of my PhD outcomes may sound like a risky invitation to be declared irrelevant or lazy by the 'pub test' of research priorities. However, like Flexner I argue that all is not lost for his so-called 'useless' research, but contend that such categorisations of utility or futility are themselves lacking in worth. Returning to Flexner's second concern about utility, he argues that utilitarian assessments of usefulness cannot fundamentally provide an exhaustive assessment of the *value* of science. Writing at a period in time heavy with shadows of the imminent darkness that beset twentieth-century history, Flexner (1939) noted:

> Curiosity, which may or may not eventuate in something useful is probably the outstanding characteristic of modern thinking...The real enemy of the human race is not the fearless and irresponsible thinker, be he right or wrong. The real enemy is the man who tries to hold the human spirit so that it will not dare to spread it wings. (p. 550)

Echoes of Flexner's arguments again persist. For example, Australian researchers Peat Leith and Holger Meinke (2015) summarise two persistent arguments of value used by contemporary science, namely (1) it produces the raw material for unknown future innovation, which will provide for future economic activity, or (2) it produces new knowledge that is inherently valuable, such as better understanding our place in the universe. I have also alluded throughout to this idea of science as intrinsically necessary. Science is not simply as a raw material for future, undiscovered innovation, but also to fulfil part of our human need to be delighted, motivated and inspired. Throughout my own life, evolving from childhood to scientist, science has provided me with essential intellectual fulfilment and connection to the world, and to other people around me. This justification for science can be readily applied to my 'useless' or insignificant PhD research. I experienced no remarkable epiphany or portent of the future during my PhD, but nonetheless I assert that my thesis research was valuable because it was *interesting*.

Reconstructing the past is inherently problematic. It is riddled with uncertainty and subject to our individual interpretations. During my PhD, I submitted a paper for publication detailing an interpretation of a proxy climate record, derived from a stalagmite that formed deep in a cave on a remote Indonesian island (Lewis et al. 2011). My large flock of co-authors had disparate views about what, in particular, this stalagmite was telling us about the climates of the past. Then, when my paper was returned from the process of peer review, seemingly in shreds, it turns out the two reviewers themselves had antithetical views about the record. In this case, the further back in time I looked, the more confused I got.

So what happens when everyone looks at the same evidence and comes up with a different explanation? How do scientists reconcile so many ideas and deliver useful outcomes? Furthermore, if scientists accept palaeoclimatology as a discipline that favours plural understandings, how helpful can it be to Australian families—those adjudicators of academic value? In reconsidering my PhD outcomes, rather than inflate the importance of my research through tenuous links to possible future climatic changes, I can instead rely on Flexner's second argument that utilitarian assessments of science do not encompass value. In this way, I can instead describe my PhD research as simply interesting.

Producing useful research—in an applied and practical sense—is an exciting prospect, but so too is producing *interesting* research. Reconstructing the past is a valuable blue-sky research pursuit, allowing us all moments of Flexner's (1939)

> striking the shackles off the human mind and setting it free for the adventures which in our day have, on the one hand, taken Hale and Rutherford

and Einstein and their peers millions upon millions of miles into the utter-
most realms of space and, on the other, loosed the boundless energy
imprisoned in the atom. (p. 548)

Palaeoclimatologists, for example, have delved into the waxing and waning of the mighty ice sheets, helped us understand human migration
patterns as our ancestors marched confidently out of Africa, shown how
abruptly and severely the climate can change and investigated fundamental
aspects of our capricious and powerful monsoon systems. Through these
endeavours, we can map out a past world where sea levels were vastly lower
and our continents were vastly bigger, where our seasons were of different
lengths and occurred at different times throughout the year, where our
ancestors could walk from Australia to Papua New Guinea and great sheets
of ice covered swathes of the Northern Hemisphere where wonderful,
expansive cities now reside.

Contributing to this scientific discipline—through blue or any other
shade research—brings great pleasure to scientists and those that read
their results. However, framing science as purposeful through either
means, either as a source of innovation or as a key means to uplifting
our collective spirits, is also limiting. I earlier detailed the enduring greatness of Edward Jenner and his work on inoculations, and I have touched
on an Ehrlich's discovery that proved useful many years afterwards. These
blue-sky wanderings are discoveries that provided both science and society
with the means to see further, higher and deeper, and gave society the
tools to solve grave problems. However, science has many other shades
that cannot be overlooked or forgotten.

In addition to enlightening us, science has a very dark side. Science and
its practitioners are equally as capable of providing malevolence and
iniquity as they are of elevation and inspiration. When mathematician
and historian Dr Jacob Bronowski stood at the threshold to Auschwitz,
he pronounced that

> We are always at the brink of the known; we always feel forward for what is
> to be hoped. Every judgment in science stands on the edge of error and is
> *personal*. Science is a tribute to what we can know although we are fallible.

Many years before, chemist Fritz Haber's monumental chemical discoveries fuelled the Holocaust by filling the gas chambers with Zyklon
B. Haber's chemical genius allowed highly productive agricultural systems

to be developed, which have fed the world through fixing nitrogen from the atmosphere through industrial processes. However, Haber also enthusiastically pursued weapons of mass destruction and used the same process for synthesising fertilisers to develop chlorine and other deadly gases, such as Zyklon B.

In previous chapters, I described science as human pursuit that cannot be readily dissociated from the humanity of scientists. While I discussed that my climate science was enhanced by a personal and subjective connection to scientific practice, equally Haber's chemical discoveries and ruinous career reveal a science imbued with ego and fallibility. The very human desires of scientist for recognition, legacy or power, or even nefarious motivations are not an exception to scientific practice; they are an *inherent* part of science as a human endeavour.

In this case, how can we possibly reconcile grand aspirations for with the realities of a discipline that is inherently human, and necessarily imbued simultaneously with both the basest and loftiest of humanity? If we view scientific knowledge as a source of inspiration, what can we derive for science from Haber's zest for Zyklon B? These murky figures of science also present obstacles to the simplistic assessment of research value. If Australia's hypothetical mums and dads determined that the contemporary equivalent of Zyklon B paid sufficient dividends to be funded by the taxpayer, would such academic research necessarily be useful and valuable?

THE PROBLEM WITH PROBLEMS

Hence, I have revealed both the inadequacy of framing science as a means to finding solutions to problems and the complexities of society's perennial desire to assess value in academic research. Framing science as a means to finding solutions to problems is essentially framing it as the 'solution' to the 'problem' of human ignorance. I have described how problem-solution approaches are applied both to individual studies (my Clarke Kent paper), and to specific disciplines (palaeoclimatology). During that professional communication workshop where I was tasked with a series of exercises to prepare an esoteric manuscript on the stationarity of El Niño teleconnections for public dissemination (Lewis and LeGrande 2015), I was being encouraged to view my study as a 'solution' to a 'problem.' Similarly, when palaeoclimatologists assert the usefulness of their disciplines as necessary for understanding future climate change, we are invoking this as a 'solution' to a 'problem.'

Though this problem-solution conceptualisation permeates the creation of scientific knowledge, I have pointed to critical limitations to this understanding of science as a mode of inquiry. In *Chapter 2*, I explored the fundamental limits to knowability where I posed the idea of causality as essentially an ungraspable concept. Later in *Chapter 6* I discussed a variety of modes of knowledge production, ultimately describing climate change as an issue that demands *more* than science. However, when scientists frame science as goal-oriented—as the way we find solutions to worldly problems—we suggest that scientific discoveries exhaustively uncover essential parts of the world already in existence. The assumption here is that eventually science will, or can, reveal the world. This is clearly not a tenable view of science, and nor of the world. Ultimately, this view of 'problems' as obstacles is in itself problematic. Science can be useful, it can be interesting, it can provide information that humbles and inspires, it can facilitate acts atrocious beyond reasonable comprehension, but it cannot 'solve problems.'

This problematising of the (problematic) idea of the problems may, at first, sound unhelpful. It would certainly be unlikely to pass the annual 'pub test' of research value favoured by conservative politicians (see Lamberts and Grant 2016). This is because framing research direction around problems and solutions is practically helpful. In *Chapter 6*, I maintained that science should honour its symbolic contract with society through enhanced openness. Explaining the practical outcomes or great wonderment generated by science is a practical means for science to attend to this relationship. More pragmatically, modern academic inquiry occurs in a constricted financial environment; government priorities are streamlined, grants are hyper-competitive and employment is precarious.

I am a well-funded researcher based at a well-funded research-intensive university, but such prestige does not insulate me from the realities of constrained funding. Recently, I had to pass around a metaphorical hat to purchase a laptop to undertake a new series of climate model experiments and analyse model data. Working in such a tight funding space, I certainly work to frame my research, however technical or abstract, as necessary for preparing society for future climatic change. This attentiveness to rationale, benefits, outcomes and significance is imperative if science is to maintain its relevance to society and fulfil this contract.

In addition, my demonstration of the difficulties of determining research value and worth may seem contradictory to my proposed spectrum of knowledge claims evaluated by utility.[3] These ideas are not opposed. When considering the narrow definition of utility based on practical outcomes, like Flexner, I argue for greater recognition of the 'usefulness of useless knowledge.' An attention to the value and outcomes of scientific research must also be balanced with the fundamental importance of curiosity in scientific inquiry. Scientific inquiry cannot be dictated entirely by value. By investing so heavily in problems, scientists work within a strict pre-supposed set of terms and parameters, which denies science its creative essence and denies scientists ourselves vast swathes of interesting research that is valuable beyond 'solutions.' Scientists essentially deny ourselves Professor Stuart Firestein's (2012) faith in uncertainty, pleasure in mystery and cultivation of doubt, and scientists deny ourselves aspects of the world that we cannot grasp and problems that we have not yet posed.

As discussed briefly in *Chapter 3* along with ideas of utility, scientists can alternatively consider problems as a creative process and instead of focusing purely on solving problems, scientists are inspired to pose them. Here, I discussed philosophers Deleuze and Guattari's framing of problems in terms of usefulness (Gaffney 2010). Through this revised framing, science does not attempt to 'solve' problems but rather it seeks to pose them. If problems are a core part of a creative process, scientists do not invest only in goal-oriented solutions, but more productively in our responses to problems. This revisiting of problems no longer requires that scientists have a fusion between problems and solutions, and between research and outcomes, but rather science considers that '...problems have no solution but must generate solutions in themselves, determining their own boundaries through a play of tension and release, conservation and expenditure, despair and elation' (Gaffney 2010, p. 154).

An apprehension of problems and a recognition the manifold motivations of scientists and outcomes of their research does not necessarily alter my everyday practice of science. For a start, I have already detailed how a problem-solution framing of research can be useful to both science and society. Hence I do not advocate for discarding such approaches entirely. Rather, I argue that attending to the relationship between science and society does not require the adoption the mum and dad test of value that employs only narrow criteria. Instead, curiosity must also be a key scientific principle.

If I return again to my workshop on training early career researchers to communicate their research findings, I can see in retrospect that rather than solving a problem, I had posed a new problem and embarked on a creative process in response to curiosity. It was a very small problem, even in the ordinary sense. It will not humble or uplift our collective human spirit. Nonetheless, beginning as a Sunday afternoon side project alongside my 'real' work as a Research Fellow, this paper sparked a creative process and catalysed an ongoing chain of curious-driven thought. Where has this led me? In a practical sense, I have produced follow-up studies in response to this paper's suggestions for further work. In an idealistic sense, I have been attentive to the ways in which I motivate my research and describe my results, I have been open to creative sparks and I have withheld judgement of worth and value of other's research.

Throughout this chapter, I have provided multiple lines of thoughts about the value of science, including as a source of information to society, as the toolset for innovation, as a spark of inspiration and aspiration and as a catalyst for scientists' thinking. While each is manifestly inadequate for assessing the purpose and worth of science, I argue that each element of science and scientific practice is threaded together by curiosity. Scientific research that produces the raw material for innovation incites curiosity. Equally, 'useful' predictions of climate risks still require a curious human response to harness such utility. Science that produces new knowledge of inherent value aims to inspire a creative response in our understanding of our world and our place within it. Finally, science has the capacity to incite curiosity within individual scientists. For a climate science, this may be as mundane as my meandering thoughts from my Sunday afternoon number crunching, or it may be as profound as a visceral connection to disciplinary content, such as the powerful photo of an Australian bushfire I described in *Chapter 6*.

In this way, **transparency** and curiosity act as essential counterweights in scientific practice. Within a volatile admixture of society's expectations (pubs, mums and dads and government research priorities) and scientific drivers (human motivations and collective trends) many shades of science emerge. However, when transparency about scientific motivations, practice and process is balanced with communal and individual curiosity, science best serves its complex contract with society. Within this balance, science is skewed towards darker shades of science fed by human fallibility and ego, nor skewed towards a washed out, pale inquiry shining only on short term practical 'solutions' to 'problems,' such as demonstrated both at the sociology conference and the scientific communication workshop.

NOTES

1. This concept was discussed briefly in *Chapter 6*
2. discussed the Heinrich events in further detail in *Chapter 4*
3. As described in *Chapter 3*.

GLOSSARY

Attribution   This is the scientific process of establishing the most likely cause for a detected climate change with some defined level of confidence.

Blue-sky research   This is research conducted without a clear goal and is primarily curiosity driven.

Curiosity   This term is used broadly to mean a commitment to perceiving the world in new ways and an openness to new connections.

Detection   This is the scientific process of demonstrating that climate has changed in some defined statistical sense, without providing a reason for that change.

Equilibrium climate sensitivity   The long-term change in the global average surface air temperature following a doubling of carbon dioxide in the atmosphere.

Heinrich events   A natural phenomenon occurring during the ice ages when large icebergs break of Northern Hemisphere glaciers and traverse the North Atlantic. The melt water from icebergs acts to disrupt oceanic and atmospheric circulation and causes large-scale climatic change.

Problems   Here I do not refer to problems in an ordinary sense. Rather, I refer to problems in terms of their capacity for usefulness as part of a creative process in research that allows us a deeper and richer connection to science.

Radiative forcing   Following the Intergovernmental Panel on Climate Change (2013), this is a measure of the influence a factor has in altering the balance of incoming and outgoing energy in the Earth-atmosphere system and is an index of the importance of the factor as a potential climate change mechanism.

Transparency   An openness in communication and accountability of scientific practice such that non-scientists can view how scientific knowledge is generated.

Uncertainty I refer to uncertainty broadly, meaning both not knowing and also how well something is known.

Useful Flexner's (1939) problematises the categorisation of only knowledge that produces practical outcomes as useful. This is a narrower definition of utility than employed in *Chapter 3* for assessing knowledge claims.

Useless knowledge This is Flexner's (1939) description of knowledge that does not aim to address practical and immediate concerns.

Water isotopes Isotopes are two or more forms of the same element that contain equal numbers of protons but different numbers of neutrons in their nuclei. Water isotopes are water molecules comprised of differing forms of oxygen and hydrogen and hence differ in mass and behaviour in the climate system.

REFERENCES

Braconnot P, Harrison SP, Otto-Bliesner BL, et al (2011) The paleoclimate modeling intercomparison project contribution to CMIP5. CLIVAR Exchanges 56:15–19.

Fenby C, Gergis J (2013) Rainfall variations in south-eastern Australia part 1: Consolidating evidence from pre-instrumental documentary sources, 1788–1860. International Journal of Climatology 33:2956–2972. doi: 10.1002/joc.3640.

Firestein S (2012) *Ignorance: How it Drives Science*. Oxford University Press, New York.

Flexner A (1939) The usefulness of useless knowledge. Harpers 544–552

Fyke J, Eby M (2012) Comment on "climate sensitivity estimated from temperature reconstructions of the last glacial maximum.". Science 337:1294–1294. doi: 10.1126/science.1221371.

Gaffney P (2010) The Force of the Virtual: Deleuze, Science, and Philosophy. University of Minnesota Press, Minneapolis.

Gergis J, Gallant AJE, Braganza K, et al (2011) On the long-term context of the 1997–2009 "Big Dry" in South-Eastern Australia: Insights from a 206-year multi-proxy rainfall reconstruction. Climatic Change 111:923–944. doi: 10.1007/s10584-011-0263-x.

Gergis J, Ashcroft L (2013) Rainfall variations in south-eastern Australia part 2: A comparison of documentary, early instrumental and palaeoclimate records, 1788–2008. International Journal of Climatology 33:2973–2987. doi: 10.1002/joc.3639.

IPCC (2013) Climate change 2013: The physical science basis. Contribution of Working Group I to the Fifth Assessment Report of the Intergovernmental

Panel on Climate Change [Stocker TF, Qin D, Plattner G-K, Tignor M, Allen SK, Boschung J, Nauels A, Xia Y, Bex V and Midgley PM (eds.)]. Cambridge University Press, Cambridge and New York, NY. 1535 pp. doi: 10.1017/CBO9781107415324.

Knutti R, Hegerl GC (2008) The equilibrium sensitivity of the Earth's temperature to radiation changes. Nature Geoscience 1:735–743.

Lamberts R, Grant WJ (2016) A pub brawl over research funding doesn't benefit any of us.pdf. In: The Conversation. https://theconversation.com/a-pub-brawl-over-research-funding-doesnt-benefit-any-of-us-64290. Accessed 19 Sep 2016.

Leith P, Meinke H (2015) Science must be relevant to society if it's to earn its keep. In: The Conversation. http://theconversation.com/science-must-be-relevant-to-society-if-its-to-earn-its-keep-40957. Accessed 11 May 2015.

Lewis SC, LeGrande AN, Kelley M, Schmidt GA (2010) Water vapour source impacts on oxygen isotope variability in tropical precipitation during Heinrich events. Climate of the Past 6:325–343. doi: 10.5194/cp-6-325-2010.

Lewis SC, Gagan MK Ayliffe LK, et al (2011) High-resolution stalagmite reconstructions of Australian–Indonesian monsoon rainfall variability during Heinrich stadial 3 and Greenland interstadial 4. Earth and Planetary Science Letters 303:133–142.

Lewis SC, LeGrande AN, Schmidt GA, Kelley M (2014) Comparison of forced ENSO-like hydrological expressions in simulations of the pre-industrial and mid-Holocene. Journal of Geophysical Research. doi: 10.1002/(ISSN)2169-8996.

Lewis SC, LeGrande AN (2015) Stability of ENSO and its tropical Pacific teleconnections over the Last Millennium. Climate of the Past 11:1347–1360. doi: 10.5194/cp-11-1347-2015.

Mervis J (2014a) U.S. Political Scientists Relieved That Coburn Language Is Gone. In: Science. http://www.sciencemag.org/news/2014/01/us-political-scientists-relieved-coburn-language-gone. Accessed 19 Sep 2016a.

Mervis J (2014b) Updated: What does it take to get your grant targeted by Congress? In: Science. http://www.sciencemag.org/news/2014/11/updated-what-does-it-take-get-your-grant-targeted-congress1/12. Accessed 10 Nov 2014b.

O'Sullivan JD, Daniels G, Percival TMP, Ostry, DI, Dean, JF (1996) Wireless LAN. US Patent 5487069 A. Issued 23 Jan 1996.

Schmidt GA (2010) Enhancing the relevance of palaeoclimate model/data comparisons for assessments of future climate change. Journal of Quaternary Science 25:79–87. doi: 10.1002/jqs.1314.

Schmittner A, Urban NM, Shakun JD, et al (2011) Climate sensitivity estimated from temperature reconstructions of the last glacial maximum. Science 334:1385–1388. doi: 10.1126/science.1203513.

Taylor KE, Stouffer RJ, Meehl GA (2012) An overview of CMIP5 and the experiment design. Bulletin of the American Meteorological Society 93:485. doi: 10.1175/BAMS-D-11-00094.1.

CHAPTER 8

An Invitation to the Challenge

Abstract Lewis addresses the shortcomings of antiquated understandings of scientific knowledge production outlined throughout this book. In response to these limitations, Lewis presents a set of shared commitments to science that allow scientists traditionally robust and testable scientific approaches but also permit transparency and curiosity to be valued. By demonstrating that these commitments allow a range of scientific practice, Lewis centres diversity as a great strength of scientific inquiry. Lewis argues that just as climate change poses challenges and opportunities to society, so too does climate science challenge science more broadly. In presenting this challenge, Lewis invites scientists to respond and discuss in order to meet society's changing needs for science in responding to contemporary global issues.

Keywords Transparency · Flexibility · Diversity · Creativity · Climate change · Commitment to science

My experiences of learning to be and practicing as a scientist challenge long-held concepts of scientific practice. I am now paradoxically and uncomfortably a scientist who isn't a **Scientist.** I offer an invitation forward from these challenges towards a new understanding of science. While science is unwaveringly critical of the world 'out there,' scientists often overlook incongruities *within* science. As a community, scientists revile

© The Author(s) 2017
S.C. Lewis, *A Changing Climate for Science*,
DOI 10.1007/978-3-319-54265-2_8

from reflexivity and actively avoid discussing our own processes of knowledge generation. However, scientists must be willing to engage in debates that critique our own processes of knowledge acquisition, both to acknowledge the realities of practicing contemporary science and to meet society's changing need for science.

In order to address the shortcomings of antiquated understandings of scientific knowledge production that I have outlined through this book, I present set of shared commitments. These commitments allow scientists our traditionally robust and testable scientific approaches but also permit us to value transparency, curiosity and flexibility as new key scientific descriptors. These descriptors occupy a conceptual **hinterland** of this new view of science. Just as climate change poses challenges and opportunities to society, so too does climate science challenge science more broadly. Are scientists up to this challenge?

A Scientist Who Isn't a Scientist

This has been my story of becoming and of being a scientist. Every scientist has their own story and their own set of circumstances that brought them to the world of science. As for me, I have always had an interest in the natural world, exploring and collecting parts of the world as a child. I loved science at school. In grade 5, my uncle David helped me make hydrogen gas using a mix of sodium hydroxide and aluminium, which we then we used to fill a huge balloon for my school's History of Flight competition. We nailed it! My balloon took off high above the planetary boundary layer and onwards, up into the troposphere. Needless to say, I won a coveted ribbon for achievement in flight.

Soon after, I was thrilled to start high school and to be let loose on the Bunsen burners I had long envied from watching American high school movies. That year, as a 12 year old, I begged a family friend, a chemist by training, to help with an extra-curricular science project. I spent months exposing identical items of clothing to different environmental conditions and then used our chemist friend's work laboratory to test the ability of the fabric to attenuate ultra violet radiation.[1] It was amazing! Months of hard work were rewarded with fame and fortune in the guise of a participation certificate carefully crafted in MS Word.

Later, as a university undergraduate student my experience of science became less wonderfully aimless as I learned science as a systematic enterprise with a common, formalised approach. After years of amateur science,

I had finally become a scientist. The last formal discussion I had about science as discipline was as an undergraduate some 10 years ago. At that time, I learned that by orthodox, self-prescribed definitions, science and its practitioners accumulate knowledge through the **scientific method**, which I describe as underpinned by the following key disciplinary concepts:

- The physical world can be known and revealed through scientific inquiry, and is hence **knowable**.
- Scientific knowledge claims gain **legitimacy** through their falsifiability.
- Scientific knowledge claims gain **credibility** through the implementation of the scientific method.
- Scientists use various elements of governance to establish **expertise**.
- Scientists and scientific knowledge claims are **objective** and not influenced by biases and opinions.

However, my practice of science is not readily identifiable as this science. I've now been a practicing scientist for several years; I spent 4 years undertaking a PhD in climate science, 4 years as a junior postdoctoral research and am now steaming my way through first external grant, which was generously awarded by the Australian Research Council. My PhD research reconstructed past changes in our complex climate system from incomplete, uncertain data sources that lend themselves wonderfully to plural interpretations. My current research aims to understand recent changes in climate using sophisticated computer models we can argue are unfalsifiable[2]. Indeed, I have never actually sought to reveal the fundamental nature of our world through careful implementation of the scientific method. I have, however, created new understandings of the world by the haphazard approach of data mining, which I affectionately call research by brute force.

If my experiences differ so vastly to the formalised enterprise I learned of as an undergraduate student, am I even a scientist? These formal attempts to define, contain, or at least constrain a shared scientific endeavour, act to exclude pseudo-science and social science. They also act to exclude my scientific practices. By applying the understanding of science's ways of knowing that I was taught as an undergraduate to myself, I am left being a scientist in contradiction, a scientist who does not practice science, and a scientist who is not a Scientist. My research falls within this grey area between how science is perceived and how science is pursued. The absurdity of this contradiction in itself demonstrates that an attempt to unite scientists and their knowledge through a singular and prescribed way of being is limited in its utility. Here, I propose a way forward for scientists and science.

A Changing Role for Science

The challenges to science and critiques of its approaches that I have made throughout might seem unnecessary, unhelpful or even unfair. At first I thought them so. I did not purposefully become a scientist who is not a Scientist. My own frustrations long simmered, at first vaguely conceived, shadows of ideas and an indistinct uneasiness about how what I do fits into science. My disquiet was then more clearly grasped but poorly communicated, a feeling that could not yet be quite held onto, and perhaps it would be dangerous to do so. Why would I willingly critique my long-loved science when a regular barrage of unsolicited emails from climate sceptics does so for me?

At the same time, science aims to be unwaveringly critical. Scientists seek to know further, farther, deeper and wider, and in doing so, scientists do not tolerate untested assumptions. But it seems to me that scientists are labouring under major, untenable assumptions, such as that contemporary science is still adequately defined by the scientific method, and that the scientific method is the only mean through which science is legitimated. Scientists are collectively content to overlook these internal assumptions.

This myopia is evidenced in my own work, which is not neatly defined as science. Such epistemological incongruities are unlooked for and are largely unseen, as scientists actively avoid discussing our own processes of knowledge acquisition, joking that 'the philosophy of science is as useful to scientists as ornithology is to birds,' in a quote usually attributed to physicist Richard Feynman. Scientists chose to avoid such confrontations and because 'It is not so much that science lacks access to questions concerning its own process, but that it actively renounces the conditions under which it would come to confront this process as an ontological question' (Gaffney 2010, p. 10).

While this conceptual disquiet might be difficult to acknowledge, it is also necessary to explore. I am, of course, a scientist. Hence exploring these challenges is necessary for alleviating my own lingering discomfort about the gulf between science and Science. More broadly, this is a book both for scientists and for science. These critiques and challenges are also necessary to loosen the restrictions imposed by the traditional concepts of objectivity and knowability, amongst others. In order for scientists to gaze further, farther, deeper and wider at the external world, then we must also apply science's critical eyes inward; reflexivity must accompany the attempts of science to explore the physical world.

Climate science is one type of science that requires greater reflexivity. In *Chapter 6*, I described climate science as a **post-normal** science; here we acknowledge that facts are uncertain and values are disputed, we recognise that science is a mode of inquiry with limits, and that there are gaps in scientific knowledge (Funtowicz and Ravetz 1991). We also understand particularly that climate change is a broad issue that benefits from many perspectives, and not just scientific knowledge. The important relationship through which science provides society with a means to understand the physical world and information for decision-making can be described as a **contract** (Hulme 2009) that binds science to society. Despite the importance of this contract, both for society and the relevancy of science, scientists have been sluggish in recognising this contract and its inherent temporal fluidity with the changing requirements of society.

In short, if science is to continue to be useful to society, scientists must confront the realities of diverse contemporary scientific, and desist in our attempts to force scientific practices into a singular, standard scientific method. So what, precisely, are these changing needs and expectations of society that must be attended to? While evidently different for different scientific disciplines, in *Chapter 6* I explored elements of these shifting expectations of climate science. I described an increasing expectation on scientists to communicate and engage publicly, to discuss their relationship with their work and to be open about scientific approaches and practices.

At first glance, scientists might find these evolving expectations intrusive, unwelcome or simply unnecessary to science. At the same time that climate scientists are encouraged to be more transparent, scientists are also increasing pressured to justify our credibility and our worthiness as recipients of government funding. In Australia, for example, climate scientists work within a culture of persistent suspicion and disrespect. Within a year of assuming office in 2013, former Prime Minister Tony Abbott's government began systematically dismantling infrastructure to tackle climate change. For the first time since 1930, Australia was governed without a climate change or science minister. Next, Abbott's government repealed the carbon pricing, abolished various climate change research, communication and financing organisations, and downplayed climate change threats at key G20 and UN climate talks. Elements of the mainstream media have also played a part in the undermining of scientific information, perpetuating the inaccurate and misguided criticism of climate research, and targeting specific scientific organisations (Lloyd 2013, 2014).

These political acts and inaccurate reports on climate changes reveal a deterioration in Hulme's contract with society. How should science respond? Attacks on science—through barrages of emails sent from trolls, through the mainstream or social media, and through Government policy-making—naturally act to make climate scientists feel more defensive. It's inevitable for climate scientists to be more insistent of our value and legitimacy when we feel like we are under attack. Indeed, the dialogue around climate science and climate change in Australia is often framed as a war; science is under attack, science must fight back; science must win the battles to win the war. Personally, I've become very defensive of science and scientists over Christmas lunch when it seemed like an old-fashioned uncle, or friend of a friend of a friend, saw little value in my scientific work.

As scientists feel increasingly under attack, we strictly maintain the legitimacy of our ways of knowing, we become more exclusive, and we rely more heavily on expert consensus. However, this natural defensive response is misplaced, and restricts scientific inquiry and prevents reparation of science's contract with society. Ultimately, science is increasingly facing the same crusade for relevancy that our social science cousins have long endured in their quest for recognition and funding. At times, our social science friends have emulated the scientific method and presented their value as 'science-like,' but on the whole the social sciences have been far more willing to seek and instigate change. They have embraced and rejected and re-embraced sweeping changes in theories, methodologies and their own contract with society.

Science is equally well equipped to confront its own ontological issues, but has not yet done so. A recent news article in *Nature* asked the provocative question, Is Science Broken? (Woolston 2015). Discussing the worrying spate of falsified publications in the life sciences, such as Diederik Stapel's vast fraudulent contributions to psychology, the article reported an online debate about problems in science and how to 'fix' them. However, science *cannot* be 'broken.' Hulme notes that 'Science is not just a way of knowing, it is a way of knowing rooted in a history, a geography and a sociology.' That is, the relationship between science and society—Hulme's contract—is in a state of perpetual flux. Science is malleable and at any moment, scientists are empowered to 'fix' science. Scientists can respond to society's changing appreciation and needs of science, which is epitomised by fields, such as climate change, which simply demand *more*.

A COMMITMENT TO SCIENCE

How can science respond to the epistemological challenges posed by disciplines such as climate science and the complex needs of society around such volatile issues as climate change? I have maintained that there is little value in a singular way of engaging with science and its ways of knowing; a diversity of practices and approaches is required within science. I speak primarily from my own experiences of and engagement with science, although my own responses to the insufficiencies of traditional scientific concepts for contemporary practice are broadly useful beyond my own appraisal of science. My own experience rejects the traditional concepts of knowability and objectivity in favour of a focus on transparency, curiosity and flexibility.

This revised assemblage of scientific concepts useful to my research coalesces around the idea of **progressiveness** in science. As Professor Carl Bereiter (1994) of the University of Toronto states, 'It is not necessary to believe that science is approaching some objective truth, but it is necessary to believe that today's knowledge is on the whole better than yesterday's.' Collectively, scientists can toss out the idea of following the scientific method, give up on objectivity and even bin a belief that the world is ultimately knowable, but participating in science requires us to believe fundamentally that scientific knowledge is progressive. The idea of progressiveness in science *cannot* be discarded. While one idea, or even many ideas, may be accepted and later be found to be lacking or be found to be false, as a greater body of work and a great collection of people, we progress. Scientists know more than we used to and we know some things better.

Progressiveness is as an essential motivating belief of science, or a necessary 'scaffolding of our thoughts' (Wittgenstein 1969). Through this scaffolding, scientists have established a framework for inquiry, whereby what we do improves knowledge, which is an attainable goal. Philosopher Richard Rorty (1990) argues that scientists do not need to commit to the scientific method, but rather are united by a core set of virtues, or **commitments**. In addition to a primary belief in progressiveness in *science*, Bereiter expands, that *scientists* have:

1. A commitment to work towards common understanding satisfactory to all.
2. A commitment to frame questions and propositions in ways that allow evidence to be brought to bear on them. The commitment is to seek out things that opposing sides will accept as evidence.
3. A commitment to expand the body of collectively valid propositions.
4. A commitment to allow any belief to be subjected to criticism if it will advance the **discourse**. This is not a **Cartesian** commitment to question everything; it is a willingness to sacrifice any belief in the interest of scientific progress.

Bereiter summarises these as commitments to mutual understanding, empirical testability, expansion and openness. These scientific commitments are not unique to science, they are not particularly special, and they are not distinct. Other disciplines and professions, as groups or individuals, adopt some or all of these commitments at times. However, these commitments represent a set of cultural practices that are adhered to as a (re)defining characteristic of our **'tribe'** of scientists (Latour and Woolgar 1979) who have commitment to progressiveness. While an individual scientist may deviate from commitments along the way, as a group, these are our cultural norms.

These commitments are also reflected, in part in the very concepts I have problematised throughout (i.e. objectivity and falsifiability). However, these commitments are subtly but *critically* more expansive then the traditional assemblage of scientific concepts. This new of scientific commitments provides an inclusive and flexible framework for scientific practice. This framework adeptly encompasses my views, as well as the antithetical experiences of scientific practice encountered by my fact-loving officemate, who we meet in *Chapter 6*.

By embracing these broad commitments, I can avoid the knotty conceptual tangles in my research that emerge from considering only science's traditional ideals and practices. My science certainly does not strictly adhere to the approaches that we would traditionally understand as 'scientific.' I produce research that is open to plural interpretations. My results are rarely the outcome of the scientific method. They are not necessarily falsifiable. I find claims of expertise solely through peer review to be lacking. I think the practice of science *is* intuitive and that the researcher is important and benefits from reflexivity.

By embracing these broad commitments, I can commit to openness, rather than be chided by colleagues for questioning my fundamental capacity for objectivity. I can undertake post-normal climate research without applying the scientific method as a simplified hypothetic-deductive approach. I can fold in and discuss knowledge claims from other disciplines and undertake a discourse, through which we each attempt to point out what we perceive, rather than debating scientific 'fact.' I can discuss my views, results and feelings without relying on elite scientific consensus or expertise. I can strive for transparency and openness, but not judge myself lacking as a scientist if I fall back on old habits and ascribe blame for extreme climate events, or if publish a paper in a commercial journal by necessity. Such actions do not renounce broader commitments.

Furthermore, my lamentation about my scientific status becomes wholly unnecessary within these commitments. In *Chapter 6,* I appropriated an idea from the social sciences and posed a new conceptual terrain for science as an epistemological 'hinterland' (Law 2004). This terrain encompasses a framework of progressiveness and commitment for science. I described my yearnings for great transparency, reflexivity, curiosity and democratisation as new descriptors of scientific practise that fill in this conceptual terrain. Within this loose framework of progressiveness in science and commitments by scientists, I also identify myself as a post-modern scientist. The bolting together of these two words seems discordant, oxymoronic, and nonsensical. It simply sounds wrong.

Nonetheless, this term is useful for apprehending and understanding my own practice of science. **Postmodernism** is a reappraisal of modern assumptions. It is a term applied to music, visual art, architecture, philosophy and social theory, but rarely the physical sciences. Indeed, they seem mutually incompatible. We can make a postmodern critique of science's **positivist** approaches, but can I be a postmodern scientist? I resolutely argue so. Postmodernism is a plural notion, which means different things to different disciplines, and hold different means for different people. That is, it does not necessarily mean the same thing to archaeology as it does to philosophy, or to an archaeologist as to a philosopher. For me, postmodern science is a reappraisal of traditional scientific concepts, which acknowledges the pivotal role of the researcher in research and focuses on new scientific descriptors of transparency, diversity and curiosity.

In presenting these concepts, I intend to advance science beyond an understanding of itself as a singular enterprise, with uniquely definable approaches. During my PhD, I shared a cavernous and dusty office with my dearest friend, who challenged and frustrated me with her contrary understandings of science and its ways of knowing. She believes that those who are strictly impassive are best able to serve science. She is committed to dispassionate and objective practice to the extent that we should be wary of the results of PhD students who are too interested in their research topics.

In contrast, I have long experienced science as impassioned. To me, the revelations of science are not just intellectual—they are visceral. Our excitement, disbelief and disgust at scientific facts are important. Indeed, I hope the wonder I felt during my own childhood at strange critters never abates, although the subject of my interests become ever more sophisticated. I reject the idea that the researcher is irrelevant or a hindrance to science, and rather I argue that the researcher is central.

Throughout this chapter, I have discussed how I imagine a more nuanced scientific practice for *myself*. Other scientists and other scientific disciplines will strongly prefer a more orthodox understanding of scientific practice and inquiry and will interpret a set of scientific commitments in a manner aligned more closely with existing scientific principals and ideals. My officemate might retain her faith in peer review through a commitment to allowing beliefs to be subjected to criticism. She might also maintain an investment in objectivity through a commitment to frame question and propositions in ways that allow evidence to bear on them.

These disparate conceptualisations of science—my seeking a passionate engagement with science and my fact-loving officemate maintaining a dispassionate detachment from science—appear antithetical beyond reconciliation. As such individual understandings are clearly incommensurable, we should not attempt to reconcile these, but instead we must revel in our differences. It is certainly not a weakness of science that those collected together and described by the *Quarterly Review* (1834) meeting as scientists have tenuously overlapping understandings of science's approach to knowing.

Rather, it is a great strength of our inquiry that we have such a diversity of opinions, that scientists span not only vastly different disciplines, fields of interest and approaches, but also span vast epistemological divides. This diversity of interpretations should be encouraged – science needs both my so-called Scientists and scientists. In this way, diversity can be added to the new set of descriptors for science. Our collective commitments to science encourage diversity in scientific practice, and allow all scientists to be grouped together as a Latour and Woolgar's (1979) tribe of scientists, without exclusion through strict definition, or discomfort through differing practice of science

Addressing the shortcomings of current key scientific concepts also permits an expansion of science, and not just a freeing of scientists. This hinterland is not simply a construct for my personal and professional comfort; it permits a positive change for science. Scientific disciplines such as climate science churn up a plethora of sticky questions about our traditional understandings of science and its ways of knowing. As a result of such methodological nuance and societal interest, scientists working in politicised arenas can find themselves to feeling defensive; I feel like I am under attack and that the legitimacy of my work is often rejected without due consideration.

As a community, scientists lurch from stridently defending our work to asking ourselves—is science broken? This fractured understanding of science is an artefact of the same cultural context that vexes climate scientists. Science is a set of practices that are shaped by their historical, organisational and social context, with every small and large decision, scientific knowledge is constructed within these contexts (Castree 2013). The role of science and scientists within society, and science's relationship with society, is in a state of perpetual flux.

I have described science as not simply a rigid, unchanging system of knowing but as a pursuit wonderfully infused with a **temporal contingency** that can be redefined at any moment, *if* scientists allow it. Scientists can choose to respond to the changing appreciation of science and its role in society that is epitomised by fields such as climate change. Residing within the epistemological hinterland I have outlined are the necessary tools for scientists embrace a new vision of science, a science that is malleable, changing in response to the current needs of society and providing new ways of looking at problems.

This revisited science is a pursuit that confidently and explicitly asks society what questions science is required answer. This is a science that is open to defending the legitimacy of our knowledge without relying on weary and antiquated defences of greater expertise. This is a science that is willing to pull down, rather than erect barriers, between our authoritative ways of knowing and those of the non-expert. This is a science open to embracing, rejecting and re-embracing the criteria we use for establishing the legitimacy of our enterprise. This is a science that welcomes its mutable relationship with society, and society's expectations of scientists.

AN INVITATION TO THE CHALLENGE

Every now and then I attend a public lecture with some variation on the title '*Climate change: are we up to the challenge?*' Typically, these are talks given by physical climate scientists with a multi-disciplinary focus, although sometimes for good measure, a social scientist or economist might also throw into the mix their opinion about our readiness for grand societal challenges. The speaker will typically give a summary of the state of our knowledge about the physical climate system, the projected impacts of climate change and a recap of possible avenues for mitigation and adaptation. I have always heard these types of talks as outwardly looking discussions of climate change, asking whether, as a society, we are up to the challenge posed by climate change. Can we navigate the complex, global and intergenerational challenges of climate change? Will we collectively be able to see climate change as a challenge, rather than perceive change only as a threat?

The most common question I am asked at science outreach events is not about whether climate change is happening, or whether we are already beyond hope. More typically, people are curious about how I *feel* about climate change and what worries me about the future. This keen public interest is exemplified by the science outreach project ('*Is this how you feel?*') I described in *Chapter 6*, which revolves around giving a voice to climate scientist so that we can describe how we *feel* about climate change.

As for me, sometimes I feel angered, frightened or saddened by climate change. In *Chapter 6*, I described my emotional reaction to a photo depicting a Tasmanian family sheltering under a pier from a fire front in 2013. Eventually, I had to remove the photo from my PowerPoint presentations, because each time I turned around to talk about the image, it would make me tearful. In Australia, we're used to dealing with a variable climate and the extremes it throws at us, but I worry that our resilience is perversely a weakness that prevents us from preparing for our hotter future.

At other times, I feel excited by the challenge and ready for change. We talk about the threat of current climate change, and it impacts on our lives and livelihoods. We talk about the threat of future climate change, and about what's at stake for us, our children and our natural environment.

Will our children's adult lives by recognisable to us, or will their world be shameful to us? These are, of course, important considerations. But these vast questions almost always induce feelings of despair, fear and hopelessness. We rarely talk about climate change as a problem that provides society with the opportunity for a creative and positive response. Climate change gives us a chance to re-imagine our future. Our re-imagining doesn't have to be the seismic shift portrayed in the 2004 movie *The Day After Tomorrow* or the bleak, unrelenting hopelessness depicted in 2014 movie *Snowpiercer*.

I am excited to imagine a future in which we have not only responded to climate change but have actively sought to cultivate the aspects of our world that we value the most. This is a future in which we have explicitly welcomed the big, messy and wicked 'problems' of today. This is a future where we have actively sought to protect and help those most vulnerable to the environmental, social and political uncertainties that will come with a changing climate.

Until recently, I lived in the huge, sprawling Australian city of Melbourne. Melbourne is a teenager, big and gangly and awkward; it hasn't quite yet realised that it's growing up rapidly. There is only a single city in the USA—New York City—that has a population larger than Melbourne's. Despite its size, Melbourne is reluctant to talk much about planning for an uncertain future. While some groups, of course, worry about developing, planning and change, locals seem to embrace readily freeways over public transport infrastructure. Even at the end of eastern Australia's severe, decade-long drought, Melbourne hesitated to discuss safeguarding its future water security with any reference to climate change. Given my lifelong love of bicycles, the future I imagine for Melbourne includes a lot of bike paths. For others, this might entail exciting, innovative industries and technologies, or perhaps new building designs and approaches to town planning that foster close communities.

Elsewhere, there is sadly is little opportunity to be grasped. In 2013, I went to a sombre talk by the piercingly eloquent then President of Kiribati, His Excellency Anote Tong. At the 2016 Pacific Climate Change Conference he stated:

> Climate Change is one of the greatest moral challenge of all times, a moral challenge that necessitates a whole new thinking...The question which concerns us most deeply is whether we will ever be able to emerge ahead of these escalating challenges.

The island home of his people is slowly but surely being engulfed by rising seas. President Tong was very clear that while the developed world prevaricates about climate action and baulk at the cost of acting, their very *lives* are at stake. It's already too late to save those low-lying Pacific Islands. As the President encourages his young people to migrate with dignity, he is essentially acknowledging the demise of an entire people and their culture.

Melbourne, unlike Kiribati, *does* have the opportunity to re-imagine itself as adaptive and resourceful, if only it is willing. This doesn't necessarily have to mean a future of giving up all the things we love about the present and want for the future: despite what some particularly conservative politicians and commentators claim, responding proactively to climate change does not inherently equate to a return to subsidence living. It is a chance to stop, think and envisage a cleaner and better-connected future.

The same question about our readiness for the challenge of climate change can be re-posed, and re-framed as inwardly looking. As scientists, are *we* are up to the challenge posed by climate change? Although forming just one small branch of science, climate science encapsulates many of the demands that scientists collectively face. It is but one example of a field of science where it is no longer tenable for us to view scientific knowledge as the objective outcome of the pursuit of indisputable facts. It is also a field of science with a complex and mutable relationship with society. The discipline prompts many questions of scientists—can we assert our relevance without relying on antiquated ideas about our knowledge claims? Can we acknowledge the usefulness of alternative forms of knowledge, without relinquishing the usefulness of our own? Can we seize this challenge and honestly confront the limitations of our ways of understanding? These conceptual demands that climate science places on science parallels the grand challenge that climate change presents to society.

I resolutely believe that science is up to the challenges presented by climate science. However, in order to respond to the challenge contemporary disciplines such as climate science pose to traditional views about the creation of scientific knowledge, scientists must acknowledge that there *is* a challenge. We must acknowledge that there are limits to scientific inquiry. We must acknowledge that there are many ways to understand science and many ways to understand the world. By posing contemporary disciplines as challenging to science, I am not making a criticism of science, climate science or climate scientists. Rather, this is an affirmation that, in the face of a grand challenge, science can change itself to be more useful, more productive and more relevant to contemporary society.

Just as climate change gives us a chance to re-imagine the future, climate science provides an opportunity to acknowledge and address the gulf between popular perception and actual practice of science. In this way, the contemporary challenges of climate change and climate science are interconnected. While industrialisation paved the way for vast scientific, technological and societal advances, it also seeded vast challenges, such as climate change. I have posed such challenges as equally exciting in the sense that they offer vast opportunities for change. However, such, opportunities cannot necessarily be grasped by the same modes of historical thinking that seeded such challenges.

I have purposefully avoided calling this chapter a conclusion. Why? Quite simply, I have not yet made any resolutions, recommendations, or indeed conclusions. Rather, I have paid attention to possibilities for science that have come to my mind through my own experience. These are presented as a starting point, a beginning, for science. My training and research is firmly centred on climate change and variability, an experience that has prompted an appraisal of science centred on climate science. Other disciplines hold different relationships with society and offer different challenges and different opportunities.

Therefore, I do not offer a conclusion, but instead extend an invitation to respond to these challenges. This is an invitation to scientists, and others, whose views and disciplines vary, and includes those whose views may not align with my own. In this sense, my invitation is to a conversation about science, not to a consensus about practice. Just as Melbourne's future is anyone's to imagine, so too is the future of science.

Furthermore, as I have espoused the value of diversity and transparency in science, I explicitly extend this invitation to the community beyond the scientist, including to the sceptic. Divergent opinions about the veracity of climate science emerge from an intricate confluence of factors, ranging from simple ignorance to belligerence. The profound ability of vast problems to catalyse feelings of fear and hopelessness that I discussed earlier also leads to apathy or scepticism. In making explicit this invitation, I do not suggest that climate scientists must endure trolling or abuse from sceptics for the sake of a hypothetical conversation. Rather, responding to a shifting contract between science and society requires more than science, and hence my hypothetical conversation requires more than scientists.

Finally, I return to the perennial question of visiting academics presenting invited public seminars during their Australian sabbaticals—'climate change, are we up to the challenge?' Just as my anxieties for life in a warmer future are assuaged by my optimism in our collective capacity to respond to the challenge, my discomfort in the realities of scientific practice are equally alleviated by scientists' collective capacity to grasp the opportunity for a new vision of science. Hence, I emphatically conclude that science is up to the challenge of adapting to a changing climate.

NOTES

1. This project was shown in *Chapter 5* as evidence of my childhood non-expert science activities (nature peeping).
2. See further discussion in *Chapter 3*.

GLOSSARY

Cartesian This relates to the doctrine of philosopher Descartes, with an emphasis on rational analysis and the physical elements of the world.

Commitments Rorty's (1990) concept of a core set of virtues that unite scientific endeavour under a shared commitment.

Contract A concept of Hulme's (2009) to describe the critical relationship between science and society, which is bound by a contract of understanding and obligation.

Discourse In the social sciences, discourse means more than argument or discussion, and describes a formal way of thinking and defines what can be said.

Expertise Credibility and knowledge in a particular area obtained by study, training or formal experience.

Knowable/knowability The capability of being known, apprehended and understood.

Legitimacy The legitimacy of scientific knowledge is assessed through its falsifiability and adherence to key scientific concepts, such as the scientific method. This approach distinguished true science from the claims of pseudo-science, which are conversely, lacking in legitimacy.

Positivist The concept that information derived from sensory experience and interpreted through rational and logical approaches is the true source of authoritative knowledge. Intuitive knowledge is rejected in favour of empirically based knowledge.

Postmodernism   Postmodernism is a reappraisal of modern assumptions. It is a twentieth century movement in various disciplines that marks a critical departure from modernism.

Post-normal science   Philosophers of science Funtowicz and Ravetz (1991) describe science as post-normal where we recognise it as a method of inquiry where facts are uncertain, values are disputed and the stakes are high.

Progressiveness   This is Bereiter's (1994) idea of the advance of knowledge as an essential, motivating belief of science; while 'It is not necessary to believe that science is approaching some objective truth, but it is necessary to believe that today's knowledge is on the whole better than yesterday's.'

Scientific method   An approach of systematic and repeated observation, measurement, experiment and the formulation, testing and modification of scientific hypotheses.

Scientist   I use the term Scientist to describe this orthodox understanding of science pursued through a singular methodology, and use this capitalisation to distinguish this narrow understanding from the broader term scientist.

Temporal contingency   I use temporal contingency to describe the intrinsic relationship of science with time and hence context. Following Hulme (2009) 'Science not only has a methodology, but it also has a history, a geography and a sociology' (p. 78) that must be considered.

Tribe   Latour and Woolgar's (1979) description of scientists as bound together by a set of practices. This is summarised by Law (2004) as 'Scientists have a culture. They have beliefs. They have practices. They work, they gossip, and they worry about the future. And, somehow or other, out of their work, their practices and their beliefs, they produce knowledge, scientific knowledge, accounts of reality' (p. 19).

REFERENCES

Anon (1834) On the connexion of the physical sciences. The Quarterly Review 51:54–59.

Bereiter C (1994) Implications of postmodernism for science, or, science as progressive discourse. Educational Psychologist 29:3–12.

Castree N (2013) Making Sense of Nature. Routledge, New York.

Funtowicz SO, Ravetz JR (1991) A new scientific methodology for global environmental issues. In: Costanaza R (ed) Ecological Economics the Science and

Management of Sustainability. Columbia University Press: New York, NY, USA, pp 137–152.

Gaffney P (2010) The Force of the Virtual: Deleuze, Science, and Philosophy. University of Minnesota Press, Minneapolis.

Hulme M (2009) Why We Disagree About Climate Change: Understanding Controversy, Inaction and Opportunity. Cambridge University Press, New York.

Latour B, Woolgar S (1979) Laboratory Life. The Social Construction of Scientific Facts. Sage Publications, Beverly Hills.

Law J (2004) After Method: Mess in Social Science Research. Routledge, London.

Lloyd G (2013) We got it wrong on warming says IPCC. In: The Australian. http://www.theaustralian.com.au/news/we-got-it-wrong-on-warming-says-ipcc/story-e6frg6n6-1226719672318. Accessed 10 Apr 2015.

Lloyd G (2014) Bureau of Meteorology "adding mistakes" with data modelling. In: The Australian. http://www.theaustralian.com.au/news/nation/bureau-of-meteorology-adding-mistakes-with-data-modelling/story-e6frg6nf-1227048187480. Accessed 9 Apr 2015.

Rorty R (1990) Objectivity, Relativism, and Truth. Cambridge University Press, Cambridge and New York.

Wittgenstein L (1969) On Certainty. Harper Torchbooks, New York.

Woolston C (2015) Online debates erupts to ask: s science broken?. Nature 518:277–277. doi: 10.1038/518277f.

Index

© The Author(s) 2017
S.C. Lewis, *A Changing Climate for Science*,
DOI 10.1007/978-3-319-54265-2

Printed in the United States
By Bookmasters